Metascience
How God Creates
and Sustains Reality

A Polemic
Frederick Koons

ISBN: 979-8-9996159-0-9

Dedicated to My Wife Ann

TABLE OF CONTENTS

PREFACE

As a boy, I often slept outdoors on warm summer nights. On those perfect nights when the sky was clear, and you could see the Milky Way, I would gaze for hours at the stars. My eyes were sharp then and able to resolve the tiniest pinpoint of light. Usually, I saw the sky as the ancients saw it, as a vast light-speckled dome hovering above the Earth, but there were moments when my imagination penetrated the celestial shell to see into the depths of the universe. One night, I was lying in the warm grass scanning a small patch of sky when my focus came to rest on a very dim star. I quivered in subtle awe as my mind contemplated the immense distance to that star. Then my gaze slipped into an adjacent patch of blackness, and I quivered with deeper awe. I did not realize it at the time, but in that moment, the infinite and the eternal reached down from that black hole of nothingness and touched my soul. Youthful exuberance transformed my mind into a conduit through which awe has flowed ever since, and no matter how mundane life became, a sense of wonder remained imbedded within my being to surface whenever the need arose.

Intense awe terrifies, so on that boyhood night I averted my eyes and refocused on the Milky Way to resolve that vaporous veil into a network of sparkling points of light too many to comprehend. I scanned the heavens and contemplated the vast number of stars, each star composed of a vast number of particles. I know not what I thought on that distant night, but ever since I could not help thinking how insignificant we humans are, immersed in an infinite number of things, existing momentarily in an immense expanse of matter, energy, space, and time. Yet, despite this physical insignificance, we are capable of a mental transcendence. In our imaginations, we can travel beyond the stars fueling our journeys with a gift called wonder. We have access to transcendent experiences such as beauty, justice, happiness, and truth.

A lifetime has passed since that wonder-inducing moment. Since then, I marveled at the fragile beauty of a butterfly drinking nectar from purple buddleia. I watched the seed of a daylily sprout grow into a tender shoot and become a thing of beauty. I comforted a sobbing child in the stillness of a dark night. I solved a differential equation. I was thrilled by the magnificence of La Bohéme. I gazed across the great prairie of the American heartland. I watched the graceful trajectory of a baseball leaving the cool light of the stadium into the warm darkness of a summer night. I experienced awe and ennui, joy and despair, peace and anxiety, a journey filled with countless *emotions*. I observed and felt these and a thousand other things, but mostly, I wondered: could God's hand be in all of this or is life nothing more than a silly statistical accident? I wondered about specific things, about deep things, about shallow things, but what I wondered about most was mystery of God and the meaning of life. Contemplating the mystery of God and finding the meaning of life became a quest.

To find the meaning of life entails finding a path through the forest of knowledge and I found a path, but not with a preordained purpose. My quest was not based on knowledge of what I was doing. One can live a meaningful life with little knowledge or with deep knowledge. Fulfillment is independent of any specific knowledge. Knowledge merely allows us to understand where we have been. I know many people who have lived happy, fulfilling lives without ever hearing about Kant, Aquinas, entropy, or any of a thousand other deep glades of knowledge; such are the people of grace. On the other hand, many brilliant people live lives of despair. The inverse of these two observations is also true; brilliant people can lead fulfilling lives and people with little knowledge can lead lives of despair. This merely verifies the obvious—that fulfillment is not dependent on the amount of knowledge one accumulates.

My path through the forest of knowledge passed through those glades inhabited by such great subjects. I did not pause long anywhere, so the many details the experts in those glades observe may be different from what I observed in making my journey. In contemplating God, and seeking the meaning of life, there is not time in a single lifetime to stop at any specific glade in the forest of knowledge to study the fauna and flora in depth. The length of the journey dictates that we smell the fragrance of the rose and relish the beauty rather than study its botany; that we feel and see the durability and sculptural form of a granite boulder rather than study its materiality; that we contemplate mystery of a new born fawn and the grace of its mother rather than study their *morphology*; that we immerse ourselves in the coolness of the brook rather than study its hydrodynamics; that we bathe in the moonlight rather than understand moonlight's physical characteristics.

Nevertheless, to understand the structure of the path through life we cannot ignore science. To understand the social forces that play such a big part in our lives, we cannot ignore history. To understand our individual role among the rest of humanity and how we should live, we cannot ignore religion. Hence, breadth–not depth–of knowledge is more conducive to understanding God's grace and to finding the meaning of life.

What do I mean when I ask, what is the meaning of life? I am not sure about others, but what I mean is this: our existence, and our presence on earth, has a purpose. If this were not so and life had no purpose, then surely, life would have no hope. Purpose and hope are what a God-oriented *paradigm* sets before us; therefore, my quest begins with the belief that there is a God, that there is a purpose that promises hope, because without hope there is no meaning. For me, it is not the hope that we win the lottery—that is a very transitory hope—the hope I seek is that of something beyond this life. The secret, then, to finding the meaning of life is to ascertain God's purpose for creating us and the nature of that promised hope.

Before we can ascertain God's purpose, we must first deal with two questions: Does God exist? and Who am I? The answer to the second depends on the answer to the first. The answer to the first is simply yes or no. We can neither prove with certainty that God exists, nor can we prove with certainty that God does not exist. To find the meaning of life I had to exam both possibilities. Science has done a wonderful job in describing the nature of objective reality, a description that is too often the basis for the argument that God does not exist. To establish a *plausible* argument for the opposite, that God does exist, I will present an alternative description of the nature of objective reality. I call it the "model of dual reality" (MDR); it does not prove God's existence with certainty but instead presents an answer to the question, **how does God exist**? Finding a plausible answer to the how question will provide a "soft" proof for the existence of God. I derive a soft proof with an explanation of the nature of reality at a deeper level than the level at which we experience, and science observes and describes.

The answer to the second question, who am I? derives from the statement, *I am a human being in the universe*. Finding the true nature of objective reality takes care of the universe. I now present an alternate view of a human being with a God-oriented view of *subjective reality*; a version that I will deal with in Parts II, III, and IV.

Although the universe contains enormous amounts of matter, due to the vast distances separating the cosmic objects, it is empty. Our energy originates from a lone star, our sun.

The sun is nothing but an average star, one of one hundred billion, in an average galaxy, which itself is one in a 100 billion galaxies. We exist on a minuscule bit of matter, circling a lonely spark of energy, amid the vastness of space for one instant in the relentless stream of time. If we were to focus on our isolation and our physical insignificance, we would be in danger of lapsing into hopelessness. Instead, we look at the stars, out into the darkness of that cosmic theater, and we are awed by its magnificence because we possess an even more magnificent mental function—*sapience*—the presence of a mind. We think. We contemplate. We wonder. We travel to and beyond the stars. We bring joy, sorrow, laughter, song, despair, hope, beauty, and love to the sterile stage on which we play out our dramas. Without us, without our thoughts, the universe is meaningless. I believe in God created the world, a world for which we can discover God's design by observation and reason. I am a believer, not a deist in that I do not believe God created the world as a watchmaker, wound it and let it run. I believe God, a personnel God, participates in man's affairs when He chooses. He creates at certain points in our history; He sustains constantly; and He answers prayers.

The goal that I have set out for myself is to compare my view with the comment view of reality while avoiding direct conflict with the observations and theories of science. I developed an alternative view by recognizing that what we experience, and science describes is a manifestation of how God creates and sustains at a deeper level of reality. Hence, the *common view* is of the *descriptive level* of reality, and *my view* is of an *explicative level* of reality.

By approaching the subject at two levels, I develop an explanation of reality without interfering with the triumphs of science; they have been magnificent. To be sure, it is magnificence that glows with the beauty of mathematics. When science excludes the immeasurable, science ignores that which gives meaning. Since meaning is the beat to which I am marching, I must ask: is there a way science can discover God at the explicative level? There is. I also believe that such a discovery would entail a major shift in the paradigm of modern science.

In Part I, I imagine how God might create and sustain objective reality by examining the nature of space, time, matter, and energy at the explicative level. In Part II, I examine how God participates in subjective reality by applying the mechanism developed in Part I to explain the nature of life, mind, and soul. In Part III, I will use the findings of Parts 1&2 to develop an analysis of human behavior. In Part IV, I will build on which I developed in the first three parts to argue that life has meaning only when we accept and live according to God's plan.

I have lived a long, happy life filled with joy and good fortune. I attribute this to the grace of God. For this I am grateful and should be satisfied to live out my remaining years in the bliss of family life; but the wonder in my soul drives me to observe, to learn, to analyze, and finally record. In this, I follow the dictum of Schopenhauer—*Primum vivere, deinde philosophe*—first one must live, and then one may philosophize.

CONVENTIONS

I am not a scientist; I am not a philosopher; I am not a theologian; and I am certainly not a writer. I am a plain old thinker, and this book is the culmination of a lifetime of thoughts. I am also a reader and in my 91 years have read 1400+ books, so I organized this book in the way I like reading them.

1. I italicize each pertinent word the first time it appears in the text to indicate that it is in the glossary.
2. For ease of reading, instead of footnotes, I use '*style notes,*' imbedded within the text, as a '*referent*' to provide an explanation for an associated italicized '*signifier*' in the text.
3. I use chevrons <**x**> to reference a pertinent citation by professionals found in the "citation" section at the end of the book.
4. I use the term '*common view*' to describe the view held by professionals: philosophers, scientists, psychologists, and others that contribute to the subject.
5. I use '*my view*' whenever I present a view different from the common view.

PART 1—Objective Reality

This book is not to find the universal truth; it is to present my personal view of how God might create and sustain reality and how it relates to the meaning of life. Most people have a point of view, some incidental, some profound; mine I believe, fits somewhere in between. Therefore, I present here a viewpoint, rather than a world view; I present what I understand the meaning of life to be by answering the 'what, how, and why' questions; and that necessitates a working understanding of science. However, an alternative purpose is as a polemic. The purpose of this *polemic* is to provide theists with plausible answers to questions posed by *materialism*, *skepticism*, and *cynicism*, questions that aim to demonstrate that God does not exist. I make no claim that what I write is science, philosophy, or theology; it is not an apologetic; it is merely an argument presented by a single wonderer.

Chapter 1—Principles

I developed this *polemic* to provide those who believe in God with a credible argument for countering the secular humanists whose main weapon is modern science. Secular humanists imply that since science completely explains reality based solely on space, matter, energy, and time, and the laws of physics, there is no need for a God. My argument, based on the premise that God exists, contends that there is a deeper reality that underlies the modern common view. I contend that God creates and sustains reality at an explicative level<1>, a deeper level than the descriptive level of reality that we experience, and science describes with a common view. Thus, my view does not conflict with science's view of reality. If we theists are to win the intellectual battle for individual human minds, we must develop a theistic view that counters the common view. It is not enough to invoke the response "God did it" at every contested observation, but it is essential that we explain "how" God might create and sustain reality and that requires the existence of a second form of reality, a duality that consists of both material and immaterial elements combined hylomorphically (a term used by Aristotle (384–322 B.C.) to imply that physical objects result from the integration of matter and form). I use the term, *hylomorphism*, to mean the integration of the physical and the psychical.

Hylomorphism describes my dualistic view of reality, that I believe is different from the current approaches that have been around since the Greeks first recognized that materialism alone is unable to explain the obvious existence of psychical elements such as consciousness, mind, *qualia*<2>, and soul.

The question of nature of reality breaks us into two camps: materialists that claim that only matter exists, and the legion of thinkers that claim that reality is dualistic. Dualism explains observations of the psychical that materialism cannot. The problem is finding a *plausible* explanation for the nature of dualism. There have been attempts and as far as I can tell no successes. I have found a simple and plausible explanation derived from the existence of two modalities of space. It is down a dualistic road I travel to find the true nature of reality. I developed a model based on spatial dualism by utilizing both modalities of space, discrete and continuous. The model is an outline of a basic mechanism, called the holonomic mechanism that operates at the explicative level, and manifests the results, as matter, time, and energy, at the descriptive level.

> Note 1: The word holonomic has several meanings, and I use it to reflect the Greek derivative "holo" meaning wholeness, and "nomos" meaning law in that the mechanism I present operates on the whole universe simultaneously according to specific algorithmic laws that are the source of the laws of physics.

The holonomic mechanism not only explains the basic nature of objective reality; it also explains those phenomena that form subjective, rational, and transcendent reality, namely: life, mind, and soul. I will discuss the holonomic mechanism in more detail in chapter 6. At this point we need only to know that the holonomic mechanism is my view of how God creates and sustains reality. My view of reality differs from the common view in that the common view: (i) believes that continuous space is the foundation of reality; (ii) uses equations to describe reality; and (iii)) uses energy as the impetus of motion. In my view God uses (i) discrete space as the foundation of reality; (ii) algorithms to describe reality, and (iii) information to describe reality.

Since no one can prove or disapprove the existence of God with certainty, I assume the existence of God as a premise to develop a coherent, and comprehensive view of how God might create and sustain at the explicative level. If what I believe about the duality of space and the holonomic mechanism is at all realistic, then what I present is a plausible explanation of **HOW** God might create and sustain reality.

I do not intend my thesis to be a proof of God's existence or a contribution to science, philosophy, or theology; it is a rebuttal directed at the materialists' implication that God does not exist and a counterargument for those that believe God does exist. Although what I have developed is a comprehensive thesis that may not be acceptable to the scientific community, it presents a plausible view as to HOW God creates and sustains reality. I use the word plausible to mean, "that which may or may not be true;" but which seems believable although there is no evidence to prove it true or not true.

The quest for the true nature of reality divides humanity into two groups: believers in God, (theists), and non-believers such as secular humanists that argue that human reason and ethics instead of religion is the basis for morality; that there is no need for God.

Secular humanists come in three flavors: *materialists, skeptics,* and *cynics.* The materialist believes that God does not exist because matter and energy alone construct reality and since the laws of physics define matter and energy, there is no reason to invoke a supernatural agency. The skeptic believes that God does not exist because there is no proof. The cynic believes that God does not exist because the presence of evil and suffering, argues against a God who is benevolent. Theists do not absolutely reject the materialists' view because materialism does not reject the existence of God per se. The theists' main objection is with *scientism,* the philosophical position that secular humanists hold. Scientism is the belief that only science can find the true nature of reality. Theists, on the other hand, believe that the universe, life, mind, and soul are the handiwork of God. Theists believe in God without reservation, that God not only created the universe, but also sustains it. I am a theist and invoke the assumption of existence of God to explain reality and its creation. We theists believe that in addition to material the immaterial also exist. Thus, the main argument that I will address pits those that believe in a dualistic reality against those that believe there is only a material reality.

A. Scientific Principles

Simply put, science is a search for the true nature of reality. We use the word science in a varied of ways. Sometimes we use it to mean the people who are engaged in producing it—the scientists. Sometimes we use it to describe a methodology—the so-called *scientific method*. Sometimes we use it to describe a specific kind of knowledge produced—a science. We often use the word in combination of all three connotations, for example, when we read that, "Science has produced a wealth of knowledge," it could mean, scientists, a method, or knowledge. It is in the third context that the word rightfully belongs; the word science should refer to a specific kind of knowledge, but that is where the controversy arises in the current argument between the evolutionists and the proponents of Intelligent Design. Evolutionists argue that only knowledge that derives from scientific methods is worthy of the term science. They argue that for a methodology to be science, it must be the result of the classical scientific method: observation, hypothesis, prediction, and verification.

The application of the classical method might label a discipline as science without necessarily producing knowledge that is scientific. For example, the methodology used by disciplines such as psychology and sociology use scientific methodology, but the result may not be science in the sense described above. Application of scientific methods does not necessarily produce science.

In the narrow sense, a hypothesis becomes science only if it is testable (or falsifiable). This is the definition that a materialist invokes whenever convenient. It implies that the hypothesis must make a verifiable prediction. Prediction and verification give a hypothesis stature and earns it textbook status. Although hypotheses can become science without being predictive through consensus of the scientific community on the basis that they are *plausible descriptions verified by logical deduction of known facts*, hence *postdictive science*. Darwinian evolution is a good example of postdictive science. I contend that physical science merely describes reality; it does not explain it. A notable exception is Einstein's use of the curvature of space to explain gravity. I contend that that there is a deeper level of reality that if discerned would explain what science describes. Such a view of deep reality, if presented as plausible and is verified by a logical deduction of the known facts can be more useful for explaining the totality of reality than science which cannot deal with the psychical. Primarily, science is created not by a method but by a set of general principles. The three general principles of science are: philosophical, foundational, and general. The philosophical principle is *logical positivism*; the foundational principle is *abstraction*; the general principle is *mathematical formulation.*

B. The Challenge

Arguing with the materialists is a daunting task because their main weapon is science, history's most successful enterprise for describing reality. There is no reason for the conflict because each operates in a different sphere of human activity. The role of science is to describe the reality of nature; the role of religion is to strengthen the moral fiber of humanity. However, science and religion, for most of history, proceeded on separate paths and were not in conflict. There were two times, however, when they did intersect: once when the Italian scientist, Galileo Galilei (1564–1642) defended the Polish scientist Nicolaus Copernicus' (1473–1543) heliocentric theory and later when the British naturalist Charles Darwin (1809–1882) presented his theory of evolution. Science, based on

3

demonstrable evidence, won the heliocentric-theory conflict easily, leaving behind only members of a flat-earth society that will not concede.

The Darwinian conflict continues because the materialist argues that the theory of evolution explains that life has evolved over time from pools of inorganic molecules; to single-celled bacteria; to multicellular organisms; through the various forms of animals; and eventually to humans. If we abstract this standard interpretation and plot the complexity of matter over time, the result is a slowly increasing straight line extending from a biological *singularity*, the basic element of life that appeared in primordial seas, to the creation of the first human.

The fossil record, evolution's evidence, does not bear out a straight-line material complexity curve. There were times in the history of the universe when material complexity made a *quantum jump*. Nevertheless, the materialist claims that although evolution does not explain the quantum jumps in the evolutionary record, when science finds an answer, it will be materialistic.

In Part II, I will identify five episodes in the history of the universe when material complexity made a quantum jumps in what appears to have been creation events; they are the creation of: the universe (cosmogenesis); the earth (geogenesis); life (abiogenesis); multicellularity (somagenesis); and the mind (psychogenesis).

The main theistic arguments designed to counter the materialistic implication of Darwinian evolution are *creationism* and *intelligent design*. Both have failed in the theist-materialist argument because neither can compete with the materialist's use of science as a weapon by offering a plausible description of how God creates and sustains reality. Creationism merely invokes scripture; intelligent design merely points out the improbability of creation without God. The challenge for theists is to counter science's materialistic view by finding a way "to explain" (not describe) how God creates and sustains reality without contradicting established science.

I authored this book to provide those who believe in God with a credible argument for countering the secular humanists whose main weapon is modern science. Secular humanists imply that since science completely explains reality based solely on space, matter, energy, and time, and the laws of physics, there is no need for a God. If we believers are to win the intellectual battle for individual human minds, we must develop a theistic view that counters the common view. It is not sufficient for we theists to invoke the response "God did it" at every contested observation but is essential that we explain "*how*" God might create and sustain reality. That requires the existence of a second form of reality, a duality that consists of material and immaterial elements combined hylomorphically, Hylomorphism describes my dualistic view of reality that is different from the current approaches that have been around since the Greeks first recognized that materialism alone is unable to explain the obvious existence of psychical elements such as consciousness, mind, qualia, and soul.

The question of nature's reality breaks us into two camps: materialists that claim that only matter exists, and the thinkers that claim that reality is dualistic. Dualism explains observations of the psychical that materialism cannot. The problem is finding a plausible explanation for the nature of dualism. There have been attempts and as far as I can tell no successes. I contend that the proper explanation derives from the existence of two modalities of space. It is down a dualistic road I travel to find the true nature of reality.

I envision a model based on spatial dualism by utilizing both modalities of space, discrete and continuous. The model describes a basic mechanism, the *holonomic mechanism,* which when operating at the explicative level, produces the results of which appear as matter, time, and energy at the descriptive level. The holonomic mechanism not only explains the basic nature of objective reality, but also those phenomena that form subjective, rational, and transcendent reality, namely: life, mind, and soul. At this point we need only to know that the holonomic mechanism is simply my view of how God creates and sustains reality. My view of reality differs from science's view in that: (i) science contends that continuous space is the foundation of reality; I contend God uses discrete space; (ii) science uses equations to describe reality; God utilizes algorithms; (iii) science uses energy as the impetus of motion; God uses information.

C. God Exists

A God-created world has a beginning; hence the universe has not existed forever. The inclusion of God in our beginning also rules out an existence that began as the result of a statistical event happening in an indescribable environment. Since God creates then surely there is an intelligent design, and He is the designer. What do we mean by the word God?

Different religions have different views by what the word God means. Since the issue that I am addressing is between *secular humanism* and theism, between the non-believer and the believer, my thesis is based on the existence of a "generic God." I imagine a generic God to be described as pure existence or as an absolute being from which the universe emerged. Or as described by French scientist Rene Descartes (1596-1650) who stated: ***"by the name God I understand a substance which is infinite, independent, all-knowing, l-powerful and by which I myself and everything else, if anything else exists, have been created."*** [Socrates and Sartre] by Samuel Stumpf—page 243

And this is the God this polemic has in mind, a transcendent power acting at the explicative level, described as God, pure existence, and absolute being. However, I personally believe in a Catholic (a Trinity) God is the architect of reality and defines the purpose and the rules by which reality functions. The transcendent God is perfect, formless, infinite, eternal, immutable, omnipresent, omnipotent, omniscient, and absolute good. This is how we theists describe God, but this description has little *forensic* power when dealing with the non-believers. To function as an effective polemic, my view must have a more direct application to reality. My approach is to start with the beginning as we know it.

Winding existence back in time, we start with an infinitesimal object called *"a singularity"* and there, we confront the first great question: *What came before the singularity?* I refer to whatever it was as the *pre-universe.*

Here is how I view the pre-universe: Consider the paradoxical statement, "Nothing exists." Suppose we assume the negative, that nothing means the absence of everything including space, then nothing means *nonexistence.* Then our paradoxical statement becomes "nonexistence exists" an obvious contradiction. This means that non-existence cannot exist and hence existence is infinite.

If we assume the positive in our paradoxical statement, "Nothing exists" then "nothing" is something that exists, and before the beginning of the universe the something that existed is pure existence and/or Absolute Being. The pre-universe must have had the characteristic of space, that existential emptiness of dimensionality. Anything greater than a point has dimensionality and thus existence. But if there is any existence, however small, there must be infinite existence because non-existence cannot exist beyond whatever exists. And it must be spatial, the simplest imaginable existence.

To eliminate the ambiguity of the word 'nothing', I use the word 'nothingness' to mean: "having the appearance of being nothing." Hence, when I use the word 'nothingness' in reference to the infinitely existent pre-universe that is not "physical," i.e., consisting of the space, matter, energy, and time, which emerged from the singularity. And it is not spatial in the sense that defines the dimension of our universe. Because universal space is finite, and nothingness is infinite and is not one of the "physical elements; then what is it?

D. Infinite Nothingness <2>

Before universal space, matter, energy, and time emerged from the pre-universe with the *big bang*, there existed *infinite nothingness*. It was from the infinite nothingness, this pure existence, that the world of our present experience emerged to form stars, planets, the moon, oceans, mountains, and a vast diversity of life, people, ideas, symphonies, poems, and other wondrous realities. And while I am creating my starting point, there is another thing that the infinite pre-universe must have been in addition to pure existence, it must have contained in its nature, the power to initiate the big bang and the blueprint for all that is possible, i.e., it is an omnipotent and infinite *realm of possibility*. It is this infinite nothingness that is the pre-universe that I call the transcendent God, the source of power that initiated the big bang.

This polemic is intended as an argument against secular humanism. I use the term "transcendent God" in reference to all forms of theism. I use the term **"theist"** in all forms of belief in God. to include all forms of God and use "Christian" when referring to my specific religion. This is what I refer to as the generic God. Before describing how the generic God creates reality, I present my personal view. As a Roman Catholic, I believe, as most Christians do, that God exists as a *Trinity* of persons: a transcendent entity, a creator, and a personal God—three persons sharing a single divine substance (infinite nothingness) as Father, Son, and Holy Spirit.

The question of how God creates and sustains reality must address how God might exist transcendently, creatively, and personally, i.e., as the Christian triune God: Father, Son, and Holy Spirit. I contemplate the nature of God's transcendence by seeking answers to questions that primarily deal with the nature of physical reality (*ontology*). I contemplate the existence of the personal God by seeking answers to questions about our individual lives (*theology*).

My description of a triune God is a religious conviction. Hence, any attempt to counter the *materialist*, the *skeptic*, and the *cynic* with my personal description of a triune God in any discussion with non-believers would be a non-starter. Therefore, I describe a credible generic God that is not easily dismissed with a hand-wave, but which can set the foundation for a later argument for the triune God. I devise the nature of a generic God that explains reality without a conflict with science. Before dealing with God, let me first deal, in the next three chapters, with science, the main weapon that secular humanists use in the battle with we theists.

E. Philosophical Principle

Logical Positivism is the philosophical principle that defines the role of science and isolates it from metaphysics and religion. However, positivism leads to scientism, the belief that science is the only path to understanding the true nature of reality. Positivism: (i). rejects the idea that reality has a purpose; (ii). rejects attempts to explain natural phenomena by an essence or a secret cause of things; (iii) rejects, without meaning, explanations not verifiable by the senses: and (iv) advocates the study of constant relationships without delving into the underlying causes.

The principles of positivism have morphed into a set of dogmas every bit as inviolable as any held by the most devout religion, namely: (i) matter is not conscious; (ii). nature is purely mechanical; (iii). reality proceeds based on fixed laws: (iv) the total amount of matter and energy is constant; (v) all psychical phenomena—mind, consciousness, qualia— emerge solely from the neurons in the brain, i.e., the brain is both necessary and sufficient; and (vi). biology needs only matter to explain its basic principles.

However, the view that material neurons of the brain produce all our mental experiences prevents science from dealing with the obvious immaterial elements of reality, such as: mind, consciousness, qualia, and thought.

Although positivism may sometimes impede science's search for truth, it has kept science focused while abstraction and mathematical formulation provide the tools of progress. My view is not science and is not metaphysics because my thesis extends beyond physics to include psychology and such disciplines dealing with psychical phenomena. I will settle for the word metascience to describe my view. Whereas observation and measurement are the two fundamental requirements of science, I propose that plausibility and comprehension is the standard by which we judge metascience, an ontological view underlying science.

F. The Foundational Principle

Abstraction. the foundational principle, is the is the use of a single term, simple phrase, or a model to represent natural phenomena whether it is observed. An abstraction allows science to describe without the need to explain use of a single term, simple phrase, or a model to represent natural phenomena whether or not it can be observed. There are three types of abstraction: analytical, descriptive, and physical.

Analytical abstraction applies to either: the cause of a specific observation or to a specific property of reality. For example, the English scientist Isaac Newton (1643–1727) observed the *motion* of two forms of matter—an apple and the earth—and as the story goes—he applied abstraction, force, as the cause and the mass, to the property of matter that caused the attraction. Hence, two abstractions force and mass, which are unobservable through anything other than the effects they produce. Gravitational force is the analytical abstraction for "the thing that causes an attraction between bodies of matter;" and mass is the analytical abstraction of the property of matter that causes such an attraction. What we have here is pure description sans explanation. What makes this abstraction so valuable is that the structure of the underlying reality is so coherent that mathematical equations are written to describe the observations. And a mathematical equation has such a powerful usefulness that it makes an explanation superfluous.

Descriptive abstraction applies to models that simplify observed phenomena. The model purports to cast specific observations in general terms. The ideal model is a mathematical equation because it not only is precise and general, but it is also predictive. Thus, when Newton wrote his second law of motion as **F=ma**, he was defining a relationship between two analytical abstractions, force and mass, and an observed property of motion, acceleration. The equation was inherently general since it applied to any force and any mass, hence was selectively comprehensive. In addition to equations, science utilizes other descriptive abstractions such as charts, graphs, tables, diagrams, and drawings. The theory of evolution is a verbal argument, a descriptive model stated in words. Descriptive abstraction formalizes observations in general; for example, the phrase "natural selection" is two words that stand in a general way for the evolution of species and thus acts as an abstraction for an extended range of observations and mechanisms. Certainly, science could not have advanced without descriptive abstraction.

Physical abstraction is the use of one's imagination to relate experienced observations to unobservable phenomena. Imagination is often the first step in the construction of a scientific theory. For example, the Danish scientist, Niels Bohr (1885–1962) imagined a planetary atom consisting of a nucleus about which electrons circled in discrete orbits. This allowed him to describe the atom mathematically even though electron orbits do not exist. The German-born physicist, Albert Einstein (1879–1955) was adept at using physical abstraction having based his epic general theory of relativity on experiences with elevators, town clocks, trains, and street cars.

The word *'field'* is a classic example of abstraction in that it is both extremely descriptive and highly explicable. The idea arose out of the study of electromagnetism during the 19th century by scientists such as the British scientists Michael Faraday (1791–1867) and the Scottish physicist, James Clerk Maxwell (1831–1879).

A field is an extensive region of space that either has innate properties or has acquired a property associated with proximate matter. In short, a field is a vibrant region of spatial activity. Fields such as the Higgs fields and quantum fields are innate; on the other hand, the proximity of matter induces gravitational, electrical, and magnetic fields. The bottom line is that a field is an abstraction that allows science to describe complex phenomena mathematically. Hence, the field concept allows scientists to generate equations and capture the high ground in any debate while still not being able to explain why a field exists or the nature of magnetism, gravity, and other phenomena. And what is it about space that allows it to configure itself in a field? Makes one wonder what other magical things space can produce such as the mind? We shall see!

Compare what Newton did with that of his immediate predecessor, the German astronomer Johannes Kepler (1571-1630), who used data gathered by the Danish astronomer Tycho Brahe (1546-1601), to model the motion of the planets. Kepler's physical abstraction was to imagine the orbit as an ellipse. He then described the nature of the orbits as a relationship between geometry and time. No force or mass were part of his equations. He used what he saw. Kepler had jumped from the first to the third level of abstraction, from physical to descriptive abstraction. On the other hand, Newton's introduction of analytical abstraction allowed science to explode in a fury of abstractions that became classical science and eventually modern science. Whereas "physical abstractions" derive from specific observations, analytic abstractions have an advantage of generality, thus providing generality to the resulting equations. Science could not have advanced without abstractions such as: mass, energy, momentum, entropy, space-time, antimatter, dark matter, dark energy, and the Higgs field; within the framework of the current scientific view

8

of objective reality, science has verified the first six in this list, the last three are still unverified.

G. The General Principle

Nothing establishes a hypothesis as a scientific theory or law as much as presenting it in the form of a mathematical equation. Equations are inherently predictive in a testable *way*. Once science evaluates and verifies the predictions of a mathematical hypothesis, it advances to the stature of a theory or a law. Once an equation becomes a theory or a law, it then becomes a tool for *mathematical formulation*, the principle that all subsequent observations must fit the proven equation. If an observation does not fit an established law, then a new law is either required or additional abstraction formulated. For example, cosmologists formulated an abstraction called dark matter to make the observations of galactic motion fit the established equations of motion. Furthermore, a mathematical equation, once applied successfully to multiple observations becomes an acceptable tool for mathematical formulation, and despite not being testable or verifiable, can apply to speculative hypotheses such as the formulation of multiverse theories.

Because of the application of scientific principles, and without an explanation of why things happen the way that they do, the scientific community has developed an extraordinary description of objective reality as illustrated by the two great accomplishments of modern science—relativity and quantum mechanics. It is the fundamental property underlying the coherence of objective reality that made the success of science possible. Scientists are adept at describing the coherence of nature but have not been successful in finding a plausible way of describing nature's inherent cohesiveness, by which I mean finding a mechanism that explains not only the relationships among the elements of objective reality—for example, gravity and the other forces—but also the relationship between the physical elements of objective reality with the psychical elements such as life, mind, and soul that are associated with subjective, rational, and transcendent reality. If science is a search for truth, it will never find it if it restricts itself to only matter and energy as the basis for all reality. Science based solely on a foundation of matter and energy can merely describe objective reality. We need something beside the abstraction *"emergent"* to explain subjective, rational, and transcendent reality; we need a plausible theory of dualism. I will suggest one in the rest of my view. But first, let me deal with where science has taken us.

Chapter 2—Science

The history of science depends on how it is defined. I defined in the last chapter as a search for the true nature of reality. This is a definition that encompasses a wide range of human endeavors. The amount of science prior to the classical period was a minimum, I will deal just with the major accomplishments in the classical and modern periods.

A. Classical Science

Classical science had an awaking near the end of the Renaissance when the Polish scientist Nicolaus Copernicus (1473-1543) replaced the prevailing geocentric theory (with the earth at the center of the solar system) by the heliocentric theory in which the planets rotated while revolving around the sun. The true beginning of classical science began with the Italian scientist Galileo in the early part of the 17th century with his studies of the motion of falling objects. Isaac Newton, continuing the work done by the Italian scientist, Brahe, and Kepler, built the foundation of classical science by applying analytical abstraction to the observations of motion and was able to describe the result with the equations of motion and gravity.

1. Mechanics—Matter in Motion

In 1606, Galileo observed and measured the motion of falling bodies and described what he found. In addition, he also realized the principle of relativity—that the laws of motion were the same in all inertial frames—those not accelerating. Although Isaac Newton based his physics on absolute time and space, he also adhered to Galileo's principle of relativity restating it precisely for mechanical systems to read: *as far as the laws of mechanics are concerned, all observers in inertial motion are equally privileged, and no preferred state of motion can be attributed to any particular inertial observer.*

In the late 17th century, Newton derived the law of gravity and four laws that described the motion of bodies and the forces acting on them: (i) his first law, the principle of inertia was the statement "*bodies in motion remained in motion and bodies at rest remained at rest, unless acted upon by a net external force*"; (ii) his second law was the relationship among applied force (F), mass (m), and acceleration (a) in the form of the equation: **F=ma**; (iii) his third law was another statement: *when an object exerts a force on a second object, the second object exerts an equal and opposite force on the first object*; (iv) His equation of gravity is:

$$F = G \times \frac{(m1 \times m2)}{r^2} \tag{1}$$

> Note 2: (F) is the force of gravity; (G) is the gravitational constant; (m) is mass; (r) is the distance of separation.

Newton developed calculus concurrently with the German Gottfried Leibniz (1646–1716) and then applied it to his equations of motion. Swiss Leonhard Euler (1707–1783), and Irish William Hamilton (1805–1865), and Italian Joseph-Louis Lagrange (1736–1813) used calculus to expand the applicability of Newton's equation to situations where it was not appropriate for: (i) very small scales, (ii) at very high speeds, (iii) in very strong gravitational fields, or (iv) for geometries other than Cartesian coordinates.

2. Thermodynamics—Heat

Thermodynamics is the branch of physics that deals with heat and its relationship to energy, radiation, and physical properties of matter. The transfer of heat occurs in three forms of motion: conduction, convection, and thermal radiation. Conduction is the transfer of heat from higher temperature to lower temperature in solids; convection is the transfer of heat by the movement of gases and liquids. Thermal radiation is the electromagnetic transfer through space, of light, from one object to another. In conduction and convection, the heat travels, with thermal radiation the heat does not travel but instead generates heat at the surface that the electromagnetic wave strikes.

Along with the duality of light and the existence of ether, the mathematical form of thermal radiation was one of the main unresolved questions still lingering at the end of the classical period. The Rayleigh-Jeans-Law—the mathematical description of thermal radiation—accurately agreed with observation of the amount of electromagnetic flux radiated from a *blackbody* as a function of low frequency light but tended toward infinity at the high frequencies. The solution of this so-called ultraviolet catastrophe was one of the events that led to modern science.

Note 3: A blackbody is a body that absorbs all incident electromagnetic radiation and emits radiation when in thermal equilibrium.

3. Optics—Light

The study of light consists of optics, the science of how it occurs, and electrodynamics, the science of its creation and nature. Historically, the study of light focused first on the optics part and its main question: *is light a wave or a particle?* Despite investigations by scientists such as Christiaan Huygens (1628–1695) who as early as 1690 proposed that light was a wave propagating through a luminiferous ether, the early theories favored a particle view, especially since Newton's *Hypothesis of Light* provided sufficient support to allow the particle view to prevail throughout the 18th century. However, in 1801, the British polymath Thomas Young (1773–1829) used a double-slit experiment to produce an interference pattern that demonstrated that light was a wave; wave theory made comeback and became the predominant view in the 19th century. During the period 1815–1825 the French physicist, Augustin-Jean Fresnel (1788–1827), through his work in diffraction, dispersion, and polarization described light as a transverse wave that propagated within an elastic medium called luminiferous ether; he made the wave theory the predominant view of scientists until the early 20th century. With the question solved in favor of the wave view, science turned to the study of electricity and magnetism.

4. Electrodynamics—The Nature of Light

In 1800, invention of the battery by the Italian physicist, Alessandra Volta (1745–1827) made possible the study of electrodynamics. In 1820, the Danish physicist, Hans Oersted (1777–1851) discovered magnetism when he observed that a steady flow of electrons in a wire caused a compass needle to react. In the same year, the French scientist, Andre-Marie Ampere (1774–1836) derived the strength of the resulting magnetic force. The observation that the motion of charged particles creates a magnetic field raised the inverse question—does magnetism induce an electron current? Scientists ran experiments using static magnetic fields without finding the answer. Then in 1831, Faraday discovered: (i) that a *moving magnetic field* induces electron current in a wire; and conversely, (ii) the motion of an electron current induces a magnetic field to form around the wire. From these experiments, Faraday conceived the *concept* of a field.

In 1864, after considerable work by many scientists Maxwell utilized what they learned to derive a set of 4-equations dealing with electricity, magnetism, and induction. The four equations were based on work done by the German, Carl Gauss (1777–1855), Faraday, and Ampere. He used the equations to show that: (i) light is a transverse wave in the same ethereal medium that is the cause of electric and magnetic phenomena, (ii) the concept of a field is not just an aid to mathematics, but also possesses innate properties, and (iii) the speed of light is a constant.

According to Maxwell's theory, all optical and electrical phenomena propagate through the ether which suggested that it should be possible to experimentally determine motion relative to the ether. However, Maxwell's theory was unsatisfactory regarding the optics of moving bodies, and while he was able to present a complete mathematical model, he was not able to provide a coherent mechanical description of the ether. Because there was a distinction between the optical and electrodynamical phenomena, it was necessary to create "ether models" for the various forms of light. Attempts to unify the models or to create a complete mechanical description were not successful.

In 1887, American scientists Albert Michelson (1852–1931) and Edward Morley (1838–1923) designed and conducted the most accurate experiment to detect the ether, and it failed. Others also attempted to detect the ether and failed. Despite this proof that ether did not exist, the wave theory maintained its place in the science of light and where E is energy, m is mass, c is the speed of light. The question as to what waved remained unanswered. Meanwhile in 1887, the German physicist Heinrich Hertz 1857–1894 demonstrated the existence of electromagnetic waves and verified Maxwell's theory. In addition, the English Mathematician Oliver Heaviside (1850–1925) and Hertz, further developed the theory and introduced modernized versions of *Maxwell's equations*. The "Maxwell-Hertz" or "Heaviside-Hertz" equations subsequently formed an important basis for the further development of electrodynamics; we still use Heaviside's mathematical notation today.

B. Modern Science

The great scientists of the 18th and 19th centuries made observations and derived relationships that define the nature of heat, light, electricity, magnetism, and most other physical phenomena. At the end of the 19th century because scientists thought they had explained all the natural phenomena known at the time they assumed physical science had ended. Main accomplishments were in the areas of electromagnetism, thermodynamics, and mechanics.

1. Special Theory of Relativity

Early in the 20th century, the German theoretical physicist Max Planck (1858–1947) quantized energy and Einstein's *special theory of relativity* resurrected the particle theory of light and allowed science to ignore the question of the ether. The special theory merged space and time into a single 4-dimensional *space-time continuum*. The theory also merged the interchangeable duality for two other elements of objective reality—mass and energy—with the famous equation:

$$E = mc^2 \tag{2}$$

12

It is not known whether the Michelson-Morley experiment influenced Einstein when he developed the special theory of relativity, but he surely knew the results. He was also aware of the ad hoc theory developed by the Dutch physicist Henri Lorentz (1843–1928) and the Irish George Fitzgerald (1851–1901) to explain the Michelson-Morley results. The Lorentz-Fitzgerald theory postulated a shortening of the length in the direction of motion that compensated for the drag of the ether. The Lorentz-Fitzgerald solution obviously did not satisfy Einstein's enormous ability to conceive physical abstractions. He was a "master of physical abstraction" and the abstraction that kept his mind focused on relativity had occurred to him when he was 16 years old and imagined travelling alongside a light beam at the same velocity and imagined it as a standing (stationary) wave. Einstein made use of Maxwell's prediction of the constancy of the speed of light when he applied the principle of relativity that states that the laws of physics are *covariant*, i.e., they are the same in all frames of reference. Because the speed of light is a constant in Maxwell's equation, then the principle of relativity demands that the speed of light is constant in all frames of reference moving with linear velocity relative to one another.

Einstein was having trouble reconciling his imagination of standing waves with Maxwell's equation that did not allow standing waves when he had another eureka moment. While riding a streetcar on this way to work, he passed a bell tower, which struck 12, and he realized that when he heard the bell, he was in the present and the bell was in the past, separated by the amount of time it took the sound to catch the moving streetcar. It dawned on him that time is not absolute; that time depended on the signal source velocity; that there is no way to judge that two-events are simultaneous; that there is no absolute time. Out of this inspiration grew the concept of the space-time continuum.

The special theory of relativity in addition to connecting time and space also connected the other two physical properties, matter, and energy. Einstein realized that as a moving body approached the speed of light, it not only gained mass it also gained energy. When applied to the energy of a moving body, relativistic effects lead to the equivalence of matter and energy.

2. General Theory of Reality

In 1907, two aspects of the special theory continued to bother Einstein because: (i) it did not account for accelerating systems; and (ii) Newton's theory of gravitation assumed that gravity acted instantaneously but Maxwell's equations implied that nothing could travel faster than light. In November 1907, he had another eureka moment when he realized that in free fall, he would be weightless, and that the acceleration of the fall would cancel out the effect of gravity. He later refined this free fall mental exercise by imagining that a person in a closed chamber (like an elevator) would feel weight only when the elevator is accelerating or is stationary in a gravitational field. The thought experiment verified for him the well-known scientific supposition that the inertial mass was equal to the gravitational mass. From this, he realized that it would be impossible to tell whether a given frame of reference in a gravitational field was accelerating or was stationary. Then he went one step further and postulated that not only was the inertial mass equal to the gravitational mass, but so were their effects. This is what he meant by "the equivalence principle. "

He also imagined that a beam of light, entering a hole at one side in an upward accelerating elevator, would not exit a hole exactly in line horizontally, but would strike the opposite wall a small distance below as though the path of light was curved. Einstein reasoned that by the equivalence principle, the gravity associated with acceleration the was the same as the gravity associated with mass, then just as the light followed a curved path in an accelerating elevator, the light path in the sun's gravitational field would also be curved. Einstein realized that mass caused the curvature of space-time instead of a gravitational force at a distance. His immediate goal was to find equations that not only described the motion of matter in curved space but also described the formation of gravitational fields in space-time as a function of mass-energy.

Although Einstein did not abandon his desire to include gravity in a general equation of relativity, beginning in 1907, he spent the next four years concentrating on his studies of light quanta. Then in 1911, he grew tired of light quanta and decided to return to the quest of finding a field theory of gravity. He decided that to find a general theory of gravity, the space-time continuum required restructuring. Euclidian geometry applied only to *flat space* therefore, he needed a geometry that dealt with curved space-time.

Note 4: In flat space, form does not change. One could take a circle anywhere within the flat space and it would remain a circle.

By 1912, Einstein realized that mathematics (mathematical formulation) was a tool for discovering the laws of physics and he would have to upgrade his mathematical skills. He contacted a friend named Marcel Grossman who did a literature search and suggested that Einstein use the non-Euclidian geometry developed by the German mathematician, Bernard Riemann (1826–1866), et.al, using a multi-dimensional vector called a *tensor*.

He decided on dual reasoning using both a physical approach and a mathematical approach using Riemannian mathematics. The physical approach required that any general theory of gravity would: (i) revert to Newton's classical equations in weak gravitational fields; (ii) preserve the conservation of energy and momentum laws; and (iii) satisfy the equivalence principle. The mathematical approach required that a general law should be covariant, i.e., that the laws of physics are the same in every frame of reference.

He proceeded throughout 1912 without success until the end of the year when he found a tensor—the *Ricci tensor*—that satisfied the mathematical requirement for general covariance and nearly satisfied all the physical requirements. He did not think he found a covariant equation that: (i), was equivalent to Newton's laws in a weak gravitational field and (ii) preserved the conservation of energy and momentum. Satisfying one of the requirements failed to satisfy the other. Therefore, he put the disappointing mathematical approach aside and switched to a physical strategy.

He slaved away for two years and found a field equation he called the *Entwurf theory* that satisfied the physical requirements but was only partially covariant. He and Grossman thought that it was the best that they could do. In 1913 Einstein and his friend, the engineer Michele Besso (1873–1955), used the Entwurf to calculate the *perihelion* precession of Mercury's orbit. They produced a value of 18 versus the measured value of 43 seconds of arc per century. Despite this setback, Einstein continued to pursue the Entwurf theory despite continuing to find problems.

Einstein's tenacity kept him struggling with the equations until October 1915 when he finally abandoned the Entwurf theory and returned to a mathematical approach. He returned to his 1912 solution that he thought was incompatible with the physical parameters and made revisions. In the middle of November, although his revisions were not complete, he decided to recalculate the orbital precession of Mercury's perihelion and was thrilled when his value of 43 arc-seconds per century matched the observed value. He also recalculated the bending by the sun of a light beam and arrived at a value of 1.7 arc-seconds, about twice the value of his original value of 0.83 arc-seconds. Measurements confirmed the 1.7 arc-second answer three years later. The result of years of labor by one of the greatest minds in the history of humankind resulted in this simple looking equation:

$$G_{\mu\nu} = \frac{8\pi G}{c^4} T_{\mu\nu}$$ (3)

This equation may look simple to anyone that has studied algebra, but there is a horrendous mathematical complexity hidden in those small Greek subscripts. They signify that the **G** and **T** are tensors and if that is not enough, $G_{\mu\nu}$, expanding the Einstein tensor, equation (1) becomes:

$$R_{\mu\nu} - \frac{1}{2}g_{\mu\nu}R = \frac{8\pi G}{c^4} T_{\mu\nu}$$ (4)

Note 5: ($R\mu\nu$) is the Ricci tensor; ($g_{\mu\nu}$) is a metric tensor; (R) is the Ricci trace scalar curvature.

The left side of the equation represents space-time; the right side represents mass-energy. The tensors, when expanded yield ten linear equations (the field equations); only an extremely small fraction of humankind can solve them. The complexity of Einstein's theory is not included in this book. What interests me is not the creation and solution of the equations, but the philosophical questions not asked. The question brought to my mind when browsing Einstein's general relativity is: how is it possible to bend continuous space that is infinitely dense at each point? It is easy to imagine the curved trajectories of orbiting satellites, but not the curvature of the space that would create the trajectories.

3. Quantum Mechanics

Whereas relativity is the science of space and time extending far into the cosmic range of size, quantum mechanics is the science of matter and energy plunging into the depths of the microcosmic world. Relativity is deterministic; quantum mechanics is probabilistic. Quantum mechanics was not the offspring of one scientist as relativity was the child of Einstein. Contributions from several European scientists in the thirty-year period from 1905 to 1935 developed QM in a series of brilliant intuition-shattering theories. In 1900, Max Planck found a mathematical solution to the problem of *black body radiation* by assuming that instead of radiating heat continuously over the entire frequency spectrum, the electron oscillators of the "plum pudding model" emitted energy **incrementally.**

Note 6: The plum pudding model described the atom as a sphere of positively charged fluid in which an indeterminate number of negatively charged electrons oscillated

Planck's solution was a radical deviation from the classical view; it proposed that matter emitted energy in packets or quanta called photons. The Planck quantum model provided an accurate mathematical description of the observed spectra of heated black bodies. Plank's quantization scheme was the first step in the development of a new view of the nature of matter and energy that became known as quantum mechanics.

In 1905, Albert Einstein, in the second of four brilliant papers, solved the mystery of the photoelectric effect when he reasoned that if—as Planck had demonstrated—the emission of electromagnetic radiation consisting of packets of energy (photons), then the radiation must travel through space and impinge on matter as particles. Einstein used photons of light and explained the photoelectric effect.

15

In 1910, physicist from New Zealand Ernest Rutherford (1871–1937) discovered the nucleus of the atom. The discovery killed the classical plum pudding model and led, instead, to a planetary model of the atom in which electrons orbited the nucleus like the planets orbit the Earth.

In 1913, Niels Bohr, using physical abstraction, proposed that electrons are restricted to specific (quantized) orbits and absorbed or emitted energy only when changing orbits. Although this view violated the laws of classical physics, it correlated with the concept of quantized energy proposed by Planck and Einstein. Bohr showed that by setting the electron orbits in specific positions, an electron changing from one to another would release or absorb energy in increments equal to the quantity proposed by Planck and Einstein. The Bohr's model, a nucleus surrounded by electrons in strictly determined orbits, explained natural phenomena such as spectral lines.

In 1918, when experimenting with *alpha particles*, Rutherford concluded that the nucleus of the hydrogen atom is a positively charged particle he named the proton.

In 1923, Bohr proposed the complementary principle. It states that objects have pairs of complimentary properties that we cannot observe simultaneously. For example, whether light is a wave, or a particle depends on how we observe it. Thus, the observer became directly involved in physics and science had a plausible, if not explicative, solution for the wave/particle duality question.

In 1923, the American physicist, Arthur Compton (1892–1962), added verification to the wave/particle duality by showing that x-rays that are normally wavelike acted like particles when bounced off electrons.

In 1924, the French physicist, Louis DeBroglie (1892–1987) completed the idea of duality when he reasoned that if a wave phenomenon such as light could act like a particle, then a particle could also act like a wave. De Broglie postulated that the energy of a particle was a function of the frequency of the wave.

In 1925, the Austrian physicist Wolfgang Pauli (1900–1958), using descriptive abstraction, proposed the *exclusion principle* that states that no two electrons in an atom can exist in the same state. The state of the electron is determined by a set of characteristic properties (quantum numbers) such as its energy, angular momentum, magnetic moment, and other measurable quantities. One of the remarkable consequences of the exclusion principle is its restriction of the properties of electrons by allowing one particle to determine the properties of another.

In 1925, two Dutch physicists Samuel Goldsmiths and George Ulenbeck postulated another quantum number called spin (a physical abstraction) based on the imagined angular momentum associated with particles. The concept of spin solved observations such as the splitting of spectral lines.

In 1926, the Austrian physicist, Erwin Schrödinger (1887–1961), followed up on the concept of associating wave characteristics with particles by deriving a *wave* equation to describe atomic events. The equation treats all bodies of matter as waves. The solution to the wave equation gives the evolution of such a wave in time and space in terms of a new quantity (ψ) that represents the amplitude of the matter wave.

In 1926, Max Born (1882–1920) found a statistical interpretation pointing out that ψ^2 represents a probability, and the electron was not a small, hard particle but was a *probability wave*, more like a cloud encircling the nucleus.

In 1927, Werner Heisenberg (1901–1976) postulated the *uncertainty principle* and put the capstone on the basic theory of quantum mechanics. The principle is mathematically simple, but of vast philosophical consequence. Heisenberg claimed that on the atomic scale, we could not measure both a particle's momentum and position with precision. If we observed (measures) a measure of the particle's position would be uncertain by amount called Planck's constant.

In 1928, the American physicists Clinton Davisson (1881–1958) and Lester Germer (1896–1971) added more verification to duality when they showed that electrons normally particle-like produced wavelike diffraction patterns when penetrating through a gold foil.

in 1928, the British physicist Paul Dirac (1902–1984) in a brilliant example of mathematical formulation, merged relativity with quantum mechanics and modeled the interaction of an electron and a photon and created *quantum electrodynamics (QED)*.

In 1931, Dirac used QED to propose the existence of the positron, a new particle with the mass of an electron but with a positive charge. One year later, the American physicist, Carl Anderson (1905–1992), observed the path of a positron while studying cosmic rays from outer space. Its discovery validated Dirac's theory and opened the door to a completely new set of anti-particles.

In 1932, the British physicist James Chadwick (1891–1974), discovered the neutron within the nucleus where it shared space with the proton. Neutrons have no charge; protons have a positive charge. Like charges repel. The force of repulsion is inversely dependent on the square of the distance. This means that the force grows as the distance separating the charges gets smaller. The infinitesimally small separation of the protons in the nucleus generates an enormous repulsive force. Chadwick abstracted the *strong force* to explain the attachment of protons in a nucleus.

In 1933, Enrico Fermi (1901–1954), proposed a theory of the *weak force* that governed radioactive decay.

In 1935, the Japanese physicist Hideka Yukawa, (1907–1981), proposed the pion a particle that carried the strong force. The idea of quantizing forces—force carried by particles—explained radioactive decay. The force governing radioactive decay became the weak force carried by W and *Z bosons.*

On the eve of World War II, in addition to the two classical forces—gravity and electromagnetism—two new forces were postulated: the *strong force* held the nucleus together, and the governed spontaneous radioactive processes such as beta decay.

At that time, the particle zoo contained a modest list of creatures: electrons, protons, neutrons, neutrinos, and anti-matter particles corresponding to each matter particle, and the force carrying particles —the photon and meson. However, matter is not that simple, and in the decades that followed the war, particle physics returned as a main topic of study. Armed with more powerful particle accelerators and better particle detectors, physicists smashed the nucleus into hundreds of pieces. The result of this collective work done on the nature of matter led to the standard model, a theory of the particles of matter and the forces that govern them. Together, relativity and quantum mechanics formed an amazing foundation that firmly established modern science, thus allowing all the *ancillary* sciences to wield it as a weapon to beat down any challenge from alternate sources of knowledge. Scientism reigns supreme and, yet there are problems with the monopoly of science as the generator of truth. Science is incomplete; the physical theories fail to explain the observations of subjective, rational, and *transcendental* reality.

Chapter 3—Incompleteness

Science is only partially true because it evolved as a description of objective reality, and is ill-equipped to deal with subjective, rational, and transcendent reality. Total reality is more than just objective reality plus the *emergent* properties of matter; there is an obvious non-objective (immaterial, psychical, spiritual) presence. To use the abstraction 'emergence' to deal with the non-objective phenomena is not explanation; it is description. Scientific description works well for objective reality but not psychical phenomena. When the secular humanist invokes scientism—the belief that science is the only path to understanding the true nature of reality—in the argument about the existence of God, then I feel compelled to point out the incompleteness of science.

I have no argument with science; my argument is with materialism, scientism, and reductionism. Materialism believes that reality consists solely of matter and energy, and that no aspect of reality depends on the non-objective elements. Scientism believes that only science can explain all aspects of reality, that materialism is sufficient and there is no need to invoke dualism. Reductionism is the belief that all observations are the result of emergent properties of matter and energy. Since materialism, scientism, and reductionism all turn to the success of science as their main forensic weapon, I argue that science is incomplete because of (i) philosophical impediments; (ii) paradigmatic mind-set; and (iii) incomprehensiveness.

A. Philosophical Impediments

Science is based on the philosophy called *logical positivism*; it eliminates any possibility of the scientific view utilizing a psychical component of reality. Restricting the study of nature solely to matter and energy strongly implies that God does not exist. I believe that true science should address all possibilities, one of which is: God exists. I also believe that a philosophical view that is more inclusive than logical positivism is necessary to find an explanation of absolute reality.

The conflict between religion and science is a charade; there is no reason for the conflict because each operates in a different sphere of human activity. The role of science is to describe the reality of nature; the role of religion is to strengthen the moral fiber of humanity. Science and religion search for truth with different modes of *insight*. Scientific insight is self-restricted to observation and measurement; religious insight restricts itself to faith. Religion is not involved with science unless non-believers use science to claim there is no God, an unprovable statement. Conversely, religion does not replace science's role in the search for truth. Science's goal is the creation of knowledge; religion's goal is the creation of good. There is a place for both, but their roles cannot be mutually exclusive. Science describes phenomena at the descriptive level; science is explained at the explicative level. It is at the explicative level that science and religion might find common ground, and it is at the explicative level that I am presenting my thesis as an explanation.

B. Paradigms

A paradigm is a set of established scientific principles, widely held opinions, theories, and beliefs that influence how most scientists think and act. A paradigm becomes a collective mindset that presumes to be the arbiter of truth that presents a barrier that any new idea must surmount to becoming part of the common view. However, a paradigm may impede but cannot prevent truth. Paradigms evolve and we can identify three historical paradigms: ancient, classical, and modern.

The ancient paradigm spans the time from the ancient Greeks to the Middle Ages. During the era of the ancient paradigm, the producer of knowledge was metaphysics, and the arbiter of truth was religion. There was no tolerance for theoretical ideas; and science degenerated from its Greek beginnings to the alchemy of the dark ages. Theology ruled, and science was dormant. God was triumphant. Natural disasters were God's punishment for man's sinning against God.

The classical paradigm began with the Renaissance, which unleashed a paradigmatic shift in science. The mindset of western culture shifted gradually over four centuries from religion to science. Although science diminished the role of religion, it was not overwhelmed, and it shared the role of arbiter of truth with science. Although science represented a new way of describing natural phenomena, it did not divorce itself from religion. Instead of God being the direct cause of motion and other phenomena, science developed the laws of physics and invented concepts to describe the dynamics of reality. Reality was mechanistic, and all outcomes could be determined precisely with the proper measurements and the laws of physics. In biology, science replaced superstition and witchcraft by discovering the mechanics of life, the role of germs, blood, nerves, and other parts of the body. Life was still the creation of God but until challenged by science by *Lamarckian* then by *Darwinian* hypotheses. In all things, intuition was the guiding rule of interpretation.

The modern paradigm began at the beginning of the 20th century with the development of quantum mechanics and relativity, and it continues today. The first tenet of the classical paradigm to die was intuition, then determinism, and finally God. Positivism replaced religion as the arbiter of truth; secularism gained ascendancy; and the modern paradigm excluded God. Physics was not the only science undergoing a revolution; its success became the foundation of all other sciences and became the basis for the paradigmatic tenet that life, mind, and soul was a product of physics. Since science is based on universally accepted system of measurements and observations, the materialist uses science to win the existence of God argument by default by applying positivism's definition that only what is observable or measurable exists. Since we cannot observe or measure God, science implies that God does not exist!

The paradigmatic mindset of scientists does not stop with physical science but has spread into biological science especially with the theory of evolution that has become a "theory of everything." All observations in related fields progress from a viewpoint of evolution. For example, through the work of the American linguist *Noah Chomsky* (1928–), human speech is not an artifact; it is a result of evolution, that there is a universal grammar. Personally, I do not argue with Chomsky's view and accept it as an inductive truth while recognizing that such moments of enlightenment are plausible (but may not be true), and such is that which strengthens scientism.

C. Incomprehensiveness

The wonder of reality is how intrinsically coherent it is. The patterns of matter interacting with matter are so rhythmic and precise that they are describable by mathematical equations. Despite science's approach that has necessarily been piecemeal, it has produced a magnificent body of knowledge. Science's discovery of the constants of nature and the laws of physics have enhanced science's confidence that they are dealing with the true nature of reality. However, since scientific discovery depends on the coherent nature of reality, I believe that science's description of objective reality should also be comprehensive instead of existing as set of disparate and disconnected theories. A diversity of scientific models includes: the big bang theory, nucleosynthesis, the nebular hypothesis, plate tectonics, relativity, quantum mechanics, string theory and other theories, hypotheses, and models. Such searches for knowledge are plausible when applied to specific phenomena associated with objective reality, but ignore subjective, rational, and transcendental reality.

Scientists are aware of the lack of comprehensiveness because one of their ongoing efforts is a search for unification. The goal of physics is to find a Lagrangian, a single equation that unites not only the four forces but also all the particles in the standard model of particle physics. So far, theoretical physicists have united the electric, weak, and strong forces and all the particles in the standard model, but they have not been able to include gravity. The single greatest hope for unification has been super-string or M-theory that has been around for over 50 years without unraveling the mathematical morass that it has become. Eventually science might find a Lagrangian that unites gravity with the other physical elements of objective reality, but what about the rest of reality? Science would still be light years away from finding unification that includes both the physical phenomena (objective reality) and the psychical phenomena (subjective, rational, and transcendent reality).

The difference between physical and the psychical phenomena is so great, scientists seem not to consider such a unification possible and are satisfied to throw the abstraction of "emergence" at the problem. A comprehensive explanation of reality requires a different approach, one that overrides physical science's inability to explain subjective reality, one that explains total reality. Because science restricts its investigations to a descriptive level, it cannot find a comprehensive mechanism that could be applicable to a far greater range of phenomena than currently possible. Science requires a large and diverse number of laws, theories, and hypotheses to describe the entire range of phenomena from the beginning of the universe to the mind of man. Although the intent of this polemic is as an argument, it also may be an original approach to the nature of reality that shows how total reality can be unified and in doing so, I override scientism. I call it metascience.

D. Metascience

Metascience is a hypothetical description of existence at the *ground of reality*. In recent years, highly regarded scientists have recognized that there must be a deeper level of reality than that currently addressed by modern physics. For example, here are examples that deviate from the modern paradigm: (i) David Bohm (1917–1992) led the way when he postulated a distinction between a deeper implicate order and an explicative level of reality; (ii) Frank Tipler (views reality as a computer simulation; (iii) Michael Talbot suggests that we live in a holographic universe; (iv) Ervin Laszlo hypothesizes that the explicative level is found at a zero-point *Akashic-field*; (v) David Deutsch (1867–1941) explains how the parallel-world hypothesis can be used to connect the theories of evolution, computation, and knowledge with quantum theory, and (vi) others are examining the possibility that information and not energy drives reality.

> Note 7: Parallel worlds hypothesis is an implication of the Many Worlds interpretation of the Schrödinger wave equation.

There are, I am sure, others looking for the explicative level of reality, a deeper and more fundamental view than the current view of modern physics. Yet, these wonderers cannot seem to extract their thinking from the modern paradigm. They think in terms of force, energy, and the continuity of space and time. Scientists have come close to the simplest assumption that the space that defines our universe is discrete—the basis for metascience—from which we can find explanations for a wide range of natural and psychical phenomena. I make no claim to know the truth; I merely present an approach that offers a possible (and I believe plausible) explanation of the explicative level of reality. I do not claim that what I present is science; it is simply an approach that aims at comprehensiveness by explaining the source of the laws of physics, the nature of space, matter, time, energy, mind, and soul with plausible coherence and aims for comprehensiveness. By coherent I mean that my approach explains rather than describes observations of reality; by comprehensive I mean my approach covers the greater range of observations that not only includes objective but also subjective, rational, and transcendental reality.

Although modern science is the result of a search for the nature of reality, it is self-restricted to a partial truth. In addition to the quantum weirdness there is also much that remains unexplained in a deeper sense. What causes gravity, mass, energy? How does space bend and expand? What was there before and what lies beyond the universe? How does expanding space 'push' material galaxies apart?

Because of the positivistic restriction of a partial truth, science as presently defined, can never find the whole truth. Thus, having traveled down the long path of positivism, scientists have found themselves in a *cul-de-sac* in which the truth about life, mind, and soul have eluded them. In addition, there is belief that physical science has reached the end of its journey. But "the end of science" was an idea that appeared at the end of the 19th century, and we saw how that did not hold up in the 20th century.

Science is not dead; I believe a new paradigm will evolve, one that allows the psychical back into the search for knowledge. The new paradigm will bring ascendancy to the intelligent design hypothesis as it diminishes the descriptive status of *Darwinism*. The new paradigm will strengthen the foundation of science without superseding the laws of physics. The hope of religion will triumph over the cynicism of secular humanism. How will this happen? I cannot be sure, but I will suggest one possibility in the rest of this text. I begin with an examination of the four basic elements of realty—space, time, matter, and energy—after which I demonstrate how my approach addresses science's unanswered questions. In chapters 4, 5, 6, and 7, I will compare my view with science's view of the four elements of reality: space, matter, energy, and time. I begin with a view of space, the physical element that is the foundation of all reality.

E. Big Bang Theory

The big bang theory is the currently acceptable scientific view of the beginning of the universe. Cosmologists contend that an infinitesimal object of unknown nature called the singularity contained all the space, time, matter, and energy that formed the entire universe. About 13.7 billion years ago, the singularity briefly expanded at an explosive rate and then continued to expand at a slower rate to form the universe in which we now find ourselves. Measured observations have verified the theory. Because none of the physical elements (space, matter, time, and energy) existed prior to the big bang, the cause of the big bang must have been associated with a psychical as in non-physical factor. Over a period of several decades, scientists developed the theory as follows:

In the 17th century, Isaac Newton passed sunlight through a prism and discovered that white light was composed of a spectrum of colors.

In the 19th century, scientists developed a method for determining the material composition of stars through spectral analysis. Using a spectroscope to diffract light from the sun results in a set of characteristic dark lines embedded in the color spectrum. The German physicist, Joseph von Fraunhofer (1787–1826) discovered the *spectral lines* in the early 1800's.

In 1859, German physicists, Gustav Kirchhoff (1824–1887) and Robert Bunsen (1811–1899) found that the frequencies of the dark spectral lines found in sunlight correspond to the frequencies of bright lines produced when chemical elements are heated and made to glow in the laboratory. They explained that the bright lines are associated with the characteristic frequency of light emitted by an activated (vaporized) element. Astronomers soon learned to attach spectroscopes to their telescopes to create spectra of the stars. Consequently, the frequency of the dark lines in stars' spectra aligned with the bright lines of known elements and allowed astronomers to determine what elements the stars contained.

In 1912, the American astronomer, Henrietta Leavitt (1868–1921) discovered the relationship between the period of luminosity and the absolute magnitude of Cepheid variable stars. This provided astronomers with a reliable way to measure galactic distances.

In 1914, the American astronomer Vesto Slipher (1875–1960) announced that characteristic spectral lines observed in the light from nebulae were not at frequencies one would expect; instead, the lines shifted toward the red end of the spectrum. This red shift in spectral lines meant that the object was moving away from the observer.

In 1916, Einstein introduced the General Theory of Relativity. When Einstein solved the equation for the universe, he found the solution predicted an expanding universe. To conform to the existing paradigm that held that the universe was static (non-expanding), he attempted to produce a solution for a static universe and eliminate the expansion by adding a "fudge factor" called the cosmological constant.

In 1922 the Russian astronomer Alexander Friedman (1888–1925) found that Einstein made a mathematical mistake and even with the cosmological constant, the relativity equation predicted an expanding universe.

In 1924, the American cosmologist Edwin Hubble (1880–1953) focused the new 100-inch Mt. Wilson telescope on individual stars in the Andromeda nebulae and measured the distance using the technique developed in 1912 by Henrietta Leavitt. The distance to the Andromeda galaxy that he found made him realize that nebulas were outside our Milky Way galaxy and that the number of galaxies in the universe was far greater distant than we thought.

In 1929, Hubble analyzed the Slipher's red shift data from very distant galaxies and noticed that the more distant a galaxy from our own, the larger the red shift. This meant that the farther away the receding galaxies were from our galaxy, the higher the velocity of recession. This could only mean that the universe was expanding.

In 1931, Roman Catholic priest and astronomer, Fr. Georges Lemaitre (1894–1966) independently using the Freidman model, realized that if the expansion was real, reversing the expansion in time meant that the universe at one time had to have been infinitely small object that he called the "primeval atom."

In 1948, the Russian scientist George Gamow (1904–1965) and his student Ralph Alpher (applied classical thermodynamics and nuclear physics to the primeval atom and predicted that the remnants of the expansion would be a background of energy measuring about 5 degrees Kelvin. In addition, the theory of nucleosynthesis also predicted the relative abundance of lithium and beryllium.

In 1948, Fred Hoyle, Thomas Gold (1920–2004) and Herman Biondi argued against the Big Bang theory's implied beginning of the universe by proposing the "Steady State Theory" that hypothesized that the expansion of the universe was the result of the continuous creation of matter. The additional space produced by the continuous creation of matter caused the universe to expand in size. Hence, the universe had no beginning, there was no creation, and the was no need for God.

In 1964, Jim Peeples (1955–) Robert Dicke (1916"1997) et.al at Princeton University, without knowledge of Gamow's prediction, offered their own prediction that there should be a detectable remnant of the big bang; the remnant has since become known as the cosmic background radiation (CBR). They were constructing an instrument to search for such a signal from space when they heard that two BTL engineers at Holmdel, NJ, Arno Penzias (1933–) and Robert Wilson (1936–2007) while setting up a radio telescope to be used for wireless communication, were troubled by noise from all parts of the sky. The Princeton group realized that the remnant of the big bang that they and Gamow's group predicted was the "noise" observed by Penzias and Wilson, thus verifying the big bang theory. Most cosmologists abandoned the Steady State theory.

In 1980, the American cosmologist Alan Guth (1957–) introduced the *inflation* theory to solve the horizon problem. Subsequently, observations (i) verified the prediction of the distribution of light elements, and also (ii) the observed distribution of galaxies in the various stages of evolution provides additional support for the validation of the Big Band theory.

The science I have discussed to this point is physical science, primarily physics. However, science has a much broader range of interests than objective reality, the only *modality* of reality that physics addresses. And it is physics that the materialist invokes when they use "science" as their weapon to debunk religion. Science is more than the physical that physics addresses, science also includes biology, anthropology, meteorology, chemistry, and others. In a more general long view, science is incomplete. This opens a path for me to follow that curtails scientism by finding an alternative description of objective reality. I begin with an alternative view of space.

Chapter 4—Space

God exists at all levels of reality, and, in fact, God is the explicative level. I wondered and asked myself the question, how is God's presence possible, how is it possible in a physical sense? I derived an answer based on the nature of physical space. We know what space is, but we cannot sense it. We cannot see, hear, feel, smell, or taste it. We can only imagine it as black emptiness. However, herein appears the counter to the materialists' argument that the only meaningful view of reality is the science of matter, time, and energy because they are the only elements of reality that are experienced. Although we cannot experience space, I will argue it is not only real, it is the foundation of all reality.

To counter the materialist's belief that God is unnecessary, I explain not that God exists; that is my basic premise. A better approach is to argue not only that "God did it," but also explain "HOW" God did it! Obviously, such an explanation is a view of reality that includes more than matter; it is a view that necessarily includes a psychical component. Fortunately, there is a way of introducing the psychical into reality. First, let me begin by examining the paradigmatic common view of space.

A. Continuous Space

Whereas Euclid_described the properties of *abstract space*, other Greek philosophers wondered and argued about *physical space*. Abstract space is described mathematically; physical space is that which forms the universe. The most famous of the Greeks that argued about the nature of space was Greek philosopher Zeno of Elea_(b. 489 B.C.) who in the 5th century B.C. presented paradoxes demonstrating that: if space and time were continuous (i.e., infinitely divisible) there could be no motion. Zeno was attempting to defend his teacher the Greek philosopher Parmenides (c. 540 B.C. −?) who contended that reality was one whole and unchanging thing. Based on intuition that space is continuous—philosophers, especially—and legions that followed resolved the paradoxes and attacked the infinite divisibility by stretching math to the limits or simply ignoring the implication. Some dismissed the paradox by simply claiming that measurements were indeterminate for space below the Planck length of 1.61×10^{-35} meters and for time below the time of 5.4×10^{-44} seconds and could be ignored.

Continuity is a tacit assumption that there is only one kind of space, and it is continuous. Throughout the centuries, legions of mathematicians, scientists, and philosophers heartily endorsed the assumption of continuous space. However, I am amazed at how many perceptive philosophers, mathematicians, and scientists missed a better meaning of *Zeno's paradoxes*. Instead of the standard interpretation: "if space is continuous, motion is impossible," I believe the inverse: "if there is motion, space cannot be continuous," to be the correct interpretation. The continuity assumption does not allow for the possibility that physical space that fills our universe is different from the abstract space of mathematics. Continuous space is infinitely dense (no gaps between numbers). The assumption that physical space is continuous is based on the *Cantor-Dedekind axiom* that states that there is a one-to-one correspondence between real numbers and points on a line.

Since the real numbers form a continuity, they represent a continuous line in space. When the German mathematician, Georg Cantor (1845–1918) discovered that there are as many points in a plane or volume as on a line, he realized that every point in 3-dimensional volume is represented by a number on the real number line. This means that the real numbers represent the continuous space that also is the abstract space of mathematics, but what about physical space that determines the dimensions of the universe?

B. Space—My View

The more we examine objective reality, the more granulated it appears. The trend in the evolving descriptions of matter and energy has trended toward granularity. Since the singularity contained all the matter, energy, time, and space, it seems probable that all would retain a common granularity. One might imagine objective reality being nothing but a 3-dimensional lattice of quarks, electrons, and photons. But why stop there, why not go to the ultimate limit of granularity, and consider a point of discrete space as the foundation of objective reality? And this is what Zeno's paradox demanded. But, if objective reality was based on discrete space, then the other orders of reality—subjective, rational, and transcendental—could be based on continuous space. Metascience assumes that physical space is discrete. To understand how this assumption might work, we must first understand some of the amazing properties of infinity.

Infinity is one of those marvelous ideas that occur only in the human mind, and it does not occur there with clarity. We can only crawl around the scaffolding, seeing only the intimate parts, not the whole of its construction. Our vision of infinity is necessarily myopic. Cantor offered this definition of infinity: *"A collection of terms is infinite when it contains as parts other collections that have just as many terms as it has."* This means there are as many even numbers as there are integers in an infinite set. Consider the 2-sets:

$$1, 2, 3, 4.... n = \aleph_0$$

$$2, 4, 6, 8...2n = \aleph_0$$

Every integer on the top line creates an even number on the bottom line by multiplying by two. Hence there are as many even numbers as there are numbers. Both lines end at $\aleph_0$, referred to as aleph-null, first of the transfinite number*s*.

Cantor solved the problem of dealing with infinity by inventing set theory to treat infinity as a number and created the so-called transfinite numbers. He proposed a way of defining transfinite numbers and used aleph ($\aleph$), the first letter of the Hebrew alphabet, to represent a sequence of transfinite numbers ($\aleph_0 \ \aleph_1 \ \aleph_2 \ ...$). Only the first two transfinite numbers, $\aleph_0$ and $\aleph_1$, are of interest here because they are the only two that relate to the duality of space. $\aleph_0$ is associated with the rational numbers (the integers and the ratio of integers); $\aleph_1$ is associated with the real numbers (integers plus irrationals and rational numbers).

In addition to the parts being equal to the whole, there are other amazing properties of the infinite. For example, Cantor showed that there are as many points on the real number line as there are points in both 2-dimensional and 3-dimensional space. Hence, it is possible to represent all the points in an infinite 3-dimensional volume by a string of real numbers on a 1-dimensional line. The same holds true for rational numbers, hence, the correspondence between points and numbers means that a string of binary numbers (0's and 1's) rational numbers defines 3-dimensional discrete space. The Bohemian mathematician, Bernard Bolzano, showed that there are as many numbers, either rational or real in any segment, however small, of the real number line as there are on the entire line. Since numbers map into points in 3-dimensional space, there are as many points in any volume of space, both discrete and continuous, as there are points in the total volume of reality. We can neither add nor subtract any number from an infinite number, for example if we subtract the infinity of rational numbers, $\aleph_0$ from the infinity of real numbers, $\aleph_1$, we will still have $\aleph_1$. In other words, $\aleph_1$ minus $\aleph_0 = \aleph_1$. This implies that in comparison to $\aleph_1$, $\aleph_0$ is nothing (0). Mathematicians have shown that what is true for the real numbers is also true for the rational numbers. Hence, we can imagine the possibility of modelling reality using two modalities of space.

According to the big bang theory, the four "physical" elements of objective reality—space, matter, energy, and time—emerged from the singularity. Space is physical; and since time is the manifestation of change; and energy is the manifestation of motion; and change and motion happened after the Planck era, *space* must be the primary element. It had to exist at the beginning, i.e., within the singularity. Hence, space is the foundation from which the other physical elements, matter, energy, and time, are derived.

Since space emerged from the singularity before matter, energy, and time, and the singularity must have existed in the space-like pre-universe, total reality must be composed of two kinds of space. Hence, spatial duality is a way to explain both property dualism and substance dualism. Property dualism is the view that two distinct properties. physical and psychical emerge from a single physical substance. Substance dualism is the view that two distinct kinds of substance produce two kinds of properties. Since discrete space must be the source of physical matter and continuous space is the source of psychical mind, then the two forms of space give rise to property dualism and from a different view, a single substance, space, gives rise to two properties matter and mind, hence substance dualism. Thus, how one views space determines whether the result is property or substance dualism.

C. Discrete Space <4>

My reason for postulating the discreteness of space is philosophical. Discrete space provides the way I explain the pervasive presence of God. There is no evidence for believing that space is discrete; it is merely a unique and plausible way to build a model that describes reality.

> Note 8: Since discrete space is the foundation of the model I am presenting, I introduce two unfamiliar terms to differentiate the two kinds of space: s-points are associated with the rational numbers and form discrete space; s-gaps are the average distance between s-points. The "s" stands for special or spatial.

On the other hand, there is also no evidence that space is continuous and in fact Zeno's paradoxes deny the continuity of space. The discrete space that forms the finite volume of the universe only appears to be continuous because the average size of an *s-gap* in discrete space is extremely small (10^{-35} meters). The finite distance between s-points in discrete space defines a finite minimum length of the s-gap; the finite length eliminates the problematic infinities that show up in physics when we assume space is continuous and infinitely divisible. There are other reasons for utilizing discrete space.

Discrete space allows us to easily understand fields; the discreteness of charge and atomic orbits; duality of particles and waves; strings; the Higgs particle; the uncertainty principle; Planck length; Planck era; the constancy of the speed of light; and, indeed, the basic nature of motion, matter, energy, and time. Discrete space provides a way to explain how space can be the foundation of matter, other than through the Higgs particle. Discrete space also answers the questions that science has to ignore, namely, if light is a wave, what waves? The obvious answer is that discrete space can wave, continuous space cannot.

The German mathematicians Bernard Riemann (1826–1866) and Nicholai Lobachevsky (1792–1856) developed systems of geometry based on the mathematical curvature of space without concern for the reality of such, and Einstein applied the curvature of space directly to reality with the general theory of relativity. And since a variety of experiments have verified relativity, from a relativistic point of view, the curvature of space is real.

Incidentally, in my view: it is easy to imagine: (i) the curvature of discrete space; (ii the expansion of space with the big bang by imagining the s-gaps expanding; and (iii) an explosive expansion of the s-gaps explaining a plausible explanation of the inflationary period of the big bang simply by doubling the size of the s-gaps doubles the size of the universe

> Note 9: Analogically, imagine water as a continuous substance and snow as a discrete substance then we can make snowballs, but we cannot make water balls just as we can imagine curving discrete space but not continuous space.

Try to imagine how continuous space composed of an infinite number of points can expand; certainly not by adding more points; you cannot add to an infinite number. Consequently, that space curves and expands is a compelling argument for the reality of discrete space.

> Note 10: Using the innate dualism of space—discrete and continuous—as the source of both the material and the immaterial, means the difference disappears, i.e., the psychical is, in a sense, physical.

Finally, my main reason for assuming space exists in two modalities is that the physical space that defines the dimension of the universe is discrete, is that it provides a way to introduce hylomorphism, a dual structure of reality, the consequence of which allows us to more easily reconcile dualities like mind/matter and substance/form. I argue that a basic duality of space leads to a more comprehensive view of reality.

D. Hylomorphism <5>

Hylomorphism is a term used by Aristotle to describe the dual nature of reality existing as form and substance. Philosophers have explained dualism in several diverse ways, but the modern common view seems to have put the issue to rest by assuming only matter participates in the creation of reality. Despite science, that issue is not at rest because consciousness, the most profound experience of the immaterial, demands a dualistic view of reality. My view of hylomorphism derives from a spatial source, infinite nothingness, which has the characteristics of continuous space. Much of what science cannot describe —such as the creation of life, mind, consciousness, qualia, and soul from nothing. Using the two modalities of space—and space is a physical element—gives my approach to dualism a "physical" and irreducible foundation from which both the material and the immaterial emerge. Consider the following: At some time after the big bang, the universe was the size of a pumpkin; since it was finite then, it is still finite because nothing can expand to infinity. Because the universe is finite, there is a boundary that divides reality into two realms. The universe is one of the realms and whatever came before and exists beyond the universe—the before/beyond—is the other. Since it is space that defines the dimensions of the universe, the boundary consists of two kinds of space, discrete space defined by the rational numbers and continuous space defined by the irrational real numbers.

Continuous space is infinitely dense; it is not formative (made into forms). There are gaps between points in discrete space, hence, it is formative. Since matter has form, it must be associated with discrete space. Therefore, the space that emerged from the singularity as "universal space" to form our universe must be discrete, and the infinite nothingness of the before/beyond from which the universe emerged and into which it continues to expand must be continuous space. Because continuous space is formless, I imagine it as *nothingness*, a word I use in the positive interpretation of the phrase "nothing exists." Nothingness is not the absence of existence; it is something that exists, in fact, nothingness is both pure existence and infinite being. Consequently, the abstract space that defines the before/beyond is "infinite nothingness." By viewing space in its two modalities, we have found a way to explain substance dualism. A dualistic approach to reality simplifies the explanation of psychical functions and faculties with more specificity. For example, we can distinguish such dualities as: continuous or discrete; material or immaterial; conceptual or perceptual; *digital* or *analog*; thinking or thought; and physical or psychical. We can describe with specificity such mental ambiguities as mind, soul, consciousness, emergence, free will, and *volition*.

E. Pneuma <6>

Infinite nothingness plays three substantial roles depending on context: as an immaterial, psychical, or spiritual substance. Infinite nothingness also appears in three other modes (or hypostases): (i) the infinite nothingness that exists within the limits of the universe as *pneuma*; (ii) the pneuma that exists within a living cell as *bios*; and (iii) the pneuma that exists within the body of multicellular organisms as *nous*. Just to make matters more complex, I refer to the extent of infinite nothingness as: (i) the pre-universe, the (ii) the before/beyond, and (iii)the realm of possibility. I identify infinite nothingness, where relevant, as: (i) the architect of objective reality; (ii) the source of *mental experiences* such as *qualia, form*, and *intellect* (iii) the impetus that sustains all reality; and (iv) the *Mind of God*.

This view of reality, in which pneuma of continuous space contains points of discrete space, is a plausible and fundamental explanation of a hylomorphic structure of total reality with extensive explicative consequences. Because of the gaps in the real number line between rational numbers that represent points in 3-dimensional space there are gaps between the points in discrete space. I refer to the points of discrete space as *s-points* and the gaps as *s-gaps*: this implies they are special points among the points of continuous space. Since the rational numbers differentiate discrete space from continuous space, and the material aspect of reality derives from discrete space, then we can use the real numbers and continuous space to represent the immaterial or psychical aspect of objective reality. There is no feasible way of determining whether physical space is discrete or continuous.

If the universal space that forms the dimensionality of the universe is discrete, there is nothing preventing us from constructing a model based on discrete space by assuming that a string of rational numbers can represent the physical space of the universe. We live in a digital universe. Rational numbers determine the s-points that form the matter that defines objective reality, and the real (irrational) numbers create the immaterial substance that defines subjective, rational, and transcendent reality. The result is a three-dimensional configuration of s-points (discrete space) immersed in continuous space that fills up the gaps. This allows us to envision reality as a holomorphic structure in which the physical and psychical co-exist. The s-points form a 3-dimensional lattice-like configuration that I call the cosmic s-frame .

F. Holism—Cosmic s-Frame <7>

Since the material universe is *coterminous* with the immaterial *pneuma*, the cosmic *s-frame* is hylomorphic; it is a single integrated 3-dimensional configuration of s-points immersed in the infinite nothingness. Hence, when I use the term universe, I mean the matter plus the discrete universal space.

> Note 11: The cosmic s-frame is a configuration of s-points, and each s-point is referenced by a rational number, so the incrementation of the cosmic s-frame is not by energy but by information. Viewed in this way, we can see why *Pythagoras* claimed reality was numbers, *Parmenides* claimed that reality was a whole that did not change; and *Zeno* although misinterpreted by Aristides really meant if there is motion, space must be discrete. The discreteness of space implies the discreteness of space-time hence time is discrete and ***increments*** rather than flows. The incrementation of time leads to a duality of time.

I use '*cosmic s-frame*' to mean the universe plus the pneuma. The discrete space that forms the universe is determined by rational numbers. The continuous space that forms the pneuma is determined by the irrational real numbers. Discrete and continuous space form separate realms that together form the cosmic s-frame. We can now envision the Totality as consisting of three realms: (i) the infinite nothingness that existed before and beyond the universe is the *realm of possibility*; (ii) the pneuma in the s-gaps that is conterminous with the expanding universe is the *realm of potentiality*; and (iii) the matter and the discrete space that forms the universe is the *realm of actuality.* Together, the realms of actuality and potentiality form the cosmic s-frame. The cosmic s-frame describes the what the ancient Greeks called the One. In the common view, the word cosmos is interchangeable with the word universe without distinction. I use the word 'universe' to describe the realm of actuality, and the word 'cosmos' to describe the combination of all three realms, possibility, probability, and actuality.

> Note 12: Another way to look at my view of reality is to think of the universe as part of the cosmic s-frame, which is part of the cosmos.

At each instant, the cosmic s-frame consists of a single configuration of discrete s-points, which are associated with a set of rational numbers. Because the rational numbers are discrete so are the s-points that form discrete space that is like a single frame of a motion picture film—a three-dimensional portrayal of reality frozen in time. There is no inherent change or motion in a single frame of a motion picture film, in the same way, the cosmic s-frame at any instant is a static cosmic configuration of s-points called, a NOW, that I represented with {N}. The hylomorphic structure of a {N} that appears as 3-D configuration of s-points at the explicative level appears as the two static elements (matter and space) of objective reality at the descriptive level.

G. Incrementation <8>

Aristotle's demonstration of what has come to be known as the *isomorphism tests* which asserts that either magnitude, time and motion are all continuous, or they are all discrete.

This implies that at the explicative level a NOW cosmic s-frame, {N}, increments to the NEXT cosmic s-frame, {n}, Because of the large magnitude of incrementation, the change appears as continuous motion at the descriptive level and is measured as energy and time, the dynamic elements of objective reality. The incrementation of reconfigured cosmic s-frames {N-n} represents a single increment of objective reality along a path of incrementation. A single mechanism operating at the explicative level creates the path of incrementation of the cosmic s-frame from the beginning of the big bang to the present. Conversely, when observed at the descriptive level (common view), several disparate mechanisms such as the big bang at the beginning, the theory of evolution near the end, and the laws of physics throughout. I contend that only a single holonomic mechanism controls and implements the incrementation of the cosmic s-frame as instructions are downloaded from: (i) possibility (the before/beyond); to (ii) potentiality (the pneuma); and to (iii) actuality (the universe).

H. Incremental Motion

Since holonomic mechanism defines how reality changes, it also necessarily determines the nature of motion. I contend that nothing moves within a single s-frame. (this is what the Greek philosopher Parmenides meant by "the universe was one whole and unchanging thing"). Instead, positions of s-points change from one s-frame to the next or another subsequent frame and that change is manifested as motion. But matter does not move through space; instead, discrete space is incrementally reconfigured.

Continuous motion through space is an illusion; it is an incremental change of space that gives the illusion of motion and Zeno knew this a couple of thousand years ago. My view of the expansion of space implies that galaxies are not moving through space; they are repositioned as space expands; expanding space is pushing the galaxies apart. The question then is: by what mechanism is space pushing on matter? At the explicative level, matter—including the galaxies that are an integral part of the cosmic s-frame—is reconfigured to increase the inter galactic distance. Galaxies do not move apart as the result of space pushing matter; they are repositioned by the expansion in the s-gap.

In addition, the apparent motion of objects—photons or s-particles—is determined by the incrementation rate, the number of times the object is reconfigured. The incrementation rate for photons is equal to the inverse of the Planck era, 10^{+43} times in a cosmological second. Each incrementation displaces a photon a Planck length, $\sim 10^{-35}$ meters. with. S-particles are displaced the same Planck length each time they are reconfigured but the incrementation time is longer (increments less often) and the number of incrementations per second and hence the speed is lower. At the explicative level, then is no continuous motion—there is only incrementation.

1. Cellular Automata

"Path of" is a term that describes a dynamic discrete model that generates a sequence of patterns. Cellular implies a regular pattern of discrete elements, for example, the digital computer screen is composed of a regular 2-dimensional array of pixels. In my view, the universe is composed of a regular (standard spacing of points) 3-dimensional array of s-points called the cosmic s-frame. The word automaton means, self-activating. Cellular automaton then refers to an algorithmic computational model in which a set of rules determines the next pattern of excited cells. Repetition of the rules creates a sequence of changing patterns.

In 1970, a computer game called LIFE by the English Mathematician John Conway (1957-2020) became popular with people working in artificial intelligence. The game has three simple rules that determine the subsequent state of each pixel based on the state of its eight adjacent neighboring pixels. The rules:

a) If for a given pixel, the number of activated (turned on) neighbor exactly two, the given pixel remains in the same state in the next generation, that is, if the pixel is *on* it stays on, and if it is *off,* it remains off.

b) If the number of activated neighbors is exactly three, the given pixel will be on in the next generation whether it is on or off in its current state.

c) If the number of activated neighbors is 0, 1, 4, 5, 6, 7, or 8, the given pixel is activated in the next generation.

When a computer is: (i) run with Conway's algorithmic rules; (ii) given an *initial pixel configuration* input, and (iii) activated, interesting patterns on the screen appear to grow, move around, destroy each other, give birth, and disappear. The path of incrementation of the patterns on the screen depends on the initial configuration. Hence, the process is deterministic because each subsequent pattern can be determined from the previous pattern and an objective configuration can be determined from a sequence of calculations. Cellular automata are "telic", meaning "purposeful as in directed toward an objective."

Cellular automata show that simple algorithms can produce extremely complex results. I contend that algorithms rather than equations would be more useful for creating reality. A typical cellular automaton pales in comparison with any mechanism that God uses to create and sustain reality. However, today movies, games, and television use algorithms to mimic reality in 2-dimensional screens. Certainly, God can do it in 3-dimensions.

The holonomic mechanism is a philosophical model that assumes how God might create and sustain reality. It is not scientific; it only purports to be a plausible counter to the materialists' use of scientism to claim that only through science can reality be known, and there is no need of God. The model of dual reality (MDR) is based: (i) on discrete universal space rather than continuous space; (ii) on an algorithm, rather than a set of mathematical equations; and (iii) on the dynamics associated with information; and (iv) an impetus provided by God rather than an abstraction called energy. The model assumes the real entire universe to be a single integrated cellular array I call the cosmic s-frame.

2. The Holonomic Mechanism

Just as computer-generated cellular automata produce the excitation of a fixed array of pixels on a computer screen, the holonomic mechanism produces the incrementation of a 3-dimensional array of s-points that forms the cosmic s-frame. It is the incremental change of the NOW configuration of the cosmic s-frame {N} to the NEXT configuration {n} that creates the dynamics of reality—energy and time. The entire universe, including all living things, increment simultaneously that is represented by the symbol {N–n}.

As described above, the three requirements that define cellular automata are: an initial configuration, algorithmic information, and an impetus. The first two requirements, the initial configuration and algorithmic information, determine the path of incrementation; the impetus increments the now configuration to the next configuration of s-points and provides the dynamics (time and energy) of reality.

3. Algorithm

The cosmic algorithm determines the position of each s-point in the subsequent cosmic s-frame as the path of incrementation is produced. It should be possible to extract the laws of physics from the resulting path created by the algorithm. The transfer of the position data is *holographic*. Just as each piece of a hologram, however small, contains the whole image found on the complete holographic plate, but at diminished quality, hence in the actualization of position data from the realm of potentiality to the realm of actuality there is diminishment in clarity surrounding each s-point thus building contingency into reality. As we will see later, contingency is the basis for the creation of free will.

> Note 13: A hypothesis called digital physics contends the laws of physics have consequences that are computable by a series of approximations on a digital computer because the universe itself is computable on a universal Turing machine.

In addition to contingency, algorithms are programmed for probability, a feature that determines, in a general way, the timing of actualization of events. Finally, an algorithm is a better fit with reality because it is dynamic in the sense that its output is sequential just as is reality. Equations are static tools that do not fit the constant change associated with reality. The static nature of equations results in the need of uncertainty and the probabilistic description of the scientific view.

When God set the initial configuration for the cosmic s-frame, with the pertinent information that, with uniform incrementation determined by the *cosmic algorithm*, produced a specific telic objective cosmic s-frame. Thus, an initial configuration was set within the singularity, which after a vast number of increments the universe took shape. God inserts a *telic configuration* in the path of incrementation to implement the major and minor creation events according to God's design. A telic configuration is a {n} that is different than that which the algorithm would have created the result of which is a change to a new objective. *Noumena* determine the localized configuration of s-points that when input in the path of incrementation will manifest at the descriptive level, the object represented by the noumenal pattern. The result is a new sequence of actualizations. This is how God creates.

> Note 14: Noumena are patterns of possibilities associated with objects or processes that have been instantiated by God from the realm of possibilities to become the potential entities that exist in the pneuma that form the realm of potentiality.

Once actualized, the new pattern (local configuration) is *instantiated* in the pneuma to provide future repetitions of the created pattern. Whenever a pattern becomes part of reality, the probability of a repetition being repeated increases. The question is: how and where is the information stored that the holonomic mechanism uses to operate objective reality? The answer is that God instantiates information from the realm of possibilities into the pneuma in the realm of potentiality right next to the laws of physics and the laws of mathematics. The honest answer is, we do not know. What we do know is that the human mind can realize selected noumena. I contend that realization of perceptual noumena— qualia, and other sensations—is innate; we learn conceptual noumena when it involves innovation, invention, and creative thinking. Truth, beauty, justice, happiness, for example, are transcendental noumena.

Note 15: I use instantiate to mean the transfer (or downloading) of a possibility from the realm of possibility to the realm of potentiality. I will identify 4 modalities: (1) f-noumena that forms the basis for the creation of objective reality; (2) p-noumena forms subjective reality; (3) c-noumena forms rational reality; and (4) t-noumena forms transcendental reality.

I. Telic Configuration

With cellular automata, the initial configuration plus the algorithm determines the path of incrementation. Nevertheless, the algorithm is deterministic; the pattern at each step of the path of incrementation can be determined. The path of incrementation is an incremental sequence of predetermined events. Specific information regarding each s-point's position in the next s-frame is determined by the cosmic algorithm. The algorithm was instantiated from the infinite nothingness (the Mind of God) and stored in the pneuma. The algorithm is stored in the pneuma in the form of a string of 1's and 0's, such that when the comic s-frame is incremented a specific path of actualization will be followed. How things happened in the past will be the way things happened in the future based on the pattern for each element in the cosmic s-frame. In this way, a path of actualization will be consistent, and the manifested observations will be coherent.

The idea that reality is governed by numbers, in this case by 1's and 0's, was foreseen by Pythagoras, who believed that numbers themselves explained the true nature of the Universe. And the string of 1's and 0's represents the form and the pneuma represents the substance in Aristotle's elements that create reality. However, based on the premise that God exists and is omniscient, not only can the pattern at each step in the incrementation of the holonomic mechanism be determined, but God can work backwards and determine a configuration that will produce a desired objective. The ability to change the path of incrementation by introducing a modification in the configuration of a {N} means the holonomic mechanism is telic and God can introduce a telic configuration to induce creation events such as abiogenesis. Telic means that the holonomic mechanism is purposeful in a sense that the objective determines the telic configuration that achieves that goal. God has as much time as needed to determine the telic configuration that achieves an objective result. God's time as we see in the next section differs with the cosmological time that we experience. God is omniscient and every part of the infinite nothingness shares in the omniscience, but with a certain amount of diminishment, hence a certain amount of probability. Of course, I am not 100% certain that this is how nature works, I am only saying that if this is the way that it is, then we can explain many of science's unanswered questions concerning subjective reality, i.e., life, mind, and soul.

J. Divine Impetus <9>

In the God exists scenario that I am presenting, the singularity resides in the infinite nothingness of the pre-universe. All four physical elements, universal space, matter, energy, and time emerged from the singularity. For example, at the descriptive level, energy—the materialist's favorite impetus—induces motion. At the explicative level, energy is a result of the incrementation of the cosmic s-frame. However, since energy emerged from the singularity, it could not have caused the big bang and hence does not increment the cosmic s-frame. The same thing that caused the big bang is the thing that causes the incrementation of the cosmic s-frame. It is the thing I call God. The impetus that creates the path of incrementation must be a psychical element, such as the cause of the big bang must be the

Mind of God. I admit that the impetus is a "God Did It" assertion, but 'God exists' is my premise and what I am doing is describing how God does it. He does it with a *divine impetus*; and He began the big bang with it.

In mammal evolution, one of the factors in the increase in complexity was an increase of *cephalization*, the concentration of the nerves in the head to form the brain. The pneuma contained in a multicellular body is called nous. Along with the increased *material complexity* associated with cephalization, the nous induces a concomitant increase of consciousness. Our bodies, as localized *morphic s-frames xxx*, are an integral part of the cosmic s-frame that is incremented by the Mind of God. All the motion in and by our bodies is due to the incrementation of the cosmic s-frame. This is especially true of spontaneous

---

Note 16: I use the term *morphic s-frame* to mean the physical body of a multicellular metazoan. When it is filled with nous, I call it a "*corpus.*"

---

behavior. Our bodies are an integral part of the cosmic s-frame, consequently, the morphic s-frame that is our body increments along with the cosmic s-frame. The divine impetus causes the incrementation of every s-point including that which creates the morphic s-frame that is our body, hence, the Mind of God causes every movement of our body as part of a general incrementation. However, how our body acts depends on the information our mind supplies.

The *divine impetus* is the source of all change. God keeps all created things incrementing and provides the impetus that sustains reality. We and all life need only to supply the information, and God supplies the impetus for all our behavior. This explains the 3rd law of thermodynamics that temperature can never reach absolute zero because there will always until God chooses not to apply impetus. However, not every s-point is reconfigured as part of each incrementation; incrementation depends on the speed of the body. Because they always move with the speed of light, only s-points that form photons change position with every incrementation.

Chapter 5—Matter

The structure of matter is hierarchal. The three ranges are microscopic, macroscopic, and cosmic. The microscopic, that ranges from the smallest particles (quarks and leptons) to the 90+ or so atoms in the periodic table, is the focus of particle physics and the science of quantum mechanics and is most concerned with the weak and strong forces. Macroscopic—ranging in size from the atom to the planets—is the focus of life and the science of biology and is most concerned with the electric force. The cosmic—ranging in size from the stars to the entire universe—is the focus of cosmology and the general theory of relativity and is most concerned with the gravitational force.

At the microscopic range, particles organize identically; quarks combine the same way in every neutron and in every proton in the universe. The universe started out on the repetitive microscopic path that eventually bifurcated and sent matter in two directions: first, toward immensity and organization at a cosmic level, and second, toward complexity and organization at a macroscopic level. The path toward immensity culminated with the universe; the path toward complexity culminated in the human brain. The path to immensity is the result of the big bang, the expansion of the universe. The path of complexity is the result of the natural properties of matter and energy. The underlying question that cries for an answer is: why? Why immensity, why complexity?

A. Path of Actualization

We live on the planet Earth, revolving about a solitary star, the sun. Together along with several other planets, a belt of asteroids, and an unknown number of comets, the sun and the Earth make up the Solar System. Pluto, the dwarf planet, is 4.5 billion miles from the sun; the next nearest star is 4.3 light years away leaving lots of empty space beyond our solar system.

> Note 17: A light year is the distance light travels in a year at 186,000 miles per second which amounts to 5,870,000,000,000 miles, or 5.87×10^{22} miles/year.

There is no way to describe the immensity of the universe; the part that we observe is 96-billion light years across. We do not know its shape, but we assume it is spherical. The universe has 10^{11} stars scattered through 10^{11} galaxies. The galaxies organize as clusters and super-clusters and as filaments and sheets that form the borders of vast expanses of empty space, some 100's of millions of light years across. We scan the nighttime sky and wonder in awe while contemplating a theater that is immense yet finite, expanding yet stable, and filled with objects, yet empty. Nevertheless, the biggest wonder and one for which cosmologists have no answer is how did this fit into the singularity, an object smaller than 10^{-35} meters in diameter, the Planck length?

B. The Path of Complexity

The drive toward complexity occurs primarily at the macroscopic level at which there are three sub-levels of organization of matter: atomic, molecular, and biological. Neutrons, protons, and electrons that form the upper bound of the microscopic level function as the fundamental building blocks that form the macroscopic level. At the atomic sub-level, neutrons, protons, and electrons organize naturally in ninety plus ways to form a rich diversity of elements bursting with emergent properties. The atomic level of organization provides the building blocks for the entire universe.

On the earth, which itself is a quantum jump in material complexity, the atomic level of organization ratchets up to produce the molecular level. Although water and the vast diversity of molecules are rich in emergent properties, the force that drove complexity did not stop at the molecular level. Under stringent conditions, the element carbon combined with other key elements such as hydrogen and oxygen to form mega-molecules with specific structures of thousands of atoms. Mega-molecules were the precursors of the next step taken by matter in its drive for greater and greater complexity. At some moment in the past, in an indeterminate amount of time, it could be as long ago as 3.6-billion years—molecular organization reached a threshold of complexity and became alive with the creation of the first cell. In a brief geological period, material complexity increased not only as the result of a complex cellular structure, but because of the evolutionary radiation of a vast increase in the diversity of the cellular structures.

At the biological level, diversity not only led to greater material complexity, but also, to emergent properties not found anywhere else in the universe. Matter turns inward upon itself and forms a living cell. Subsequently, cells form extremely complex organisms culminating in the most complex form of matter in the universe, the human brain, a material object that investigates itself and the world around it. Matter evolved from the microscopic to the cosmic on the path to immensity while it evolved from the atomic to the animate on the path to complexity.

C. The Granulation of Matter

The study of matter's nature began with the ancient Greeks in the 5th century B.C. The Greek philosopher *Democritus* (c. 460–370 B.C.) theorized that matter consisted of indivisible atoms. His theory did not catch on because it ran counter to intuition. Aristotle presented a more plausible explanation when he proposed that matter was composed of four basic and smoothly *amorphous* substances found in fire, water, air, and earth. This descriptive level of matter lasted from the ancient paradigm well into the classical paradigm.

In the late 18th century, scientists realized that matter consisted of smaller constituents called elements. Since the "element" as the basis for matter was indivisible by ordinary means, the English scientist, John Dalton (1766–1844), in the early 19th century, resurrected Democritus' atomic theory. Dalton proposed a "billiard ball model" in that all matter is composed of a limited number of elements, that all elements are composed of indivisible, identical atoms, and that chemical reactions were the result of interaction of atoms and groups of atoms.

In 1897, the British physicist J.J. Thomson (1856–1940) discovered the electron. This sub-atomic particle was much smaller than the atom and obviously originated from within matter. This observation was troublesome and demanded a modification of Dalton's billiard-ball model of matter. Thomson called his new model, "plum pudding model."

With the introduction of quantum mechanics early in the 20th century, the granulation of matter advanced from the plum pudding model to the Bohr energy-level model consisting of electrons orbiting a nucleus consisting of protons and neutrons. Quantum physics coasted for a couple of decades until after World War II, when the final granulation of matter resulted with the development of the standard model.

D. The Standard Model

The common view arrived at a plateau of granulation with the standard model of particle physics when it reduced all matter to twelve particles: 6-quarks, 3-leptons, and 3-bosons. The quarks are the components of the protons and neutrons; the leptons include the electron and two other obscure particles. The bosons are the force carrying particles: the photon sustains the electromagnetic force; the gluon sustains the strong force; and gauge bosons sustain the weak force. At higher energy, there are hundreds of particles in the zoo. The equation for the standard model is a mathematical nightmare; physicists trained in quantum mechanics claim that it predicts two other undiscovered bosons, the graviton that sustains gravity and the Higgs particle that generates mass.

The standard model is a conceptual model that replaces Bohr's model. The quantum mechanical model is a collection of quarks that form the nucleus surrounded by an atomic volume filled with a *probability density function* (PDF). The PDF calculates probabilities of the electron's location that are highest at distances corresponding to Bohr's orbits. I am not sure that the last statement holds water, its 40-years since my last quantum mechanics course that the professor started with, "...*we will be studying the Schrödinger Wave equation that does a wonderful job of explaining all of the elements up to and including hydrogen!*"

How that probability density cloud confers materiality is beyond me, but it makes no difference to my view that there does not seem much in the scientific view that explains how hydrogen and all the rest of the elements acquire materiality from something called a Higgs boson. But the Higgs mechanism is not the only observation of science's that suggests the relationship between matter and space.

E. Creation from Space

(i) The big bang emerged from the singularity—an infinitesimal object of unknown substance—and expanded to form the immense universe in which we live. If the singularity was a single, infinitesimal object, and space, energy, matter, and time emerged from it, what was the nature of the substance that formed the singularity? Since energy and time are manifestations of change and matter was the last element to appear after the initial big bang, the singularity could not have been composed of matter, energy, or time. This means space: (i) was present before the first incremental change; (ii) was the substance that formed the singularity; (iii) was the substance that formed the universe; and (iv) was the substance from which matter emerged.

(ii) According to Einstein, the gravitational field is due to the curvature of space near large masses. When the space warp reaches sufficient energy, such as is happening near black holes, electron-positron pairs form, and create mass from empty space. Although science provides no explanation about the nature of strings, they certainly must be spatial.

(iii) The hydrogen atom is an example for the creation of matter from space. Science has no explanation for how hydrogen acquires materiality from a structure consisting of a single point particle called the electron that determines the volume that is 30,000 times the dimension of a proton that sits at the atoms center. Obviously, it is the emptiness of space that creates materiality.

F. The Higgs Boson

In 1964, the British physicist, Peter Higgs (1929—2024), proposed a mechanism to explain how the W and Z gauge bosons acquire the mass that makes them different from photons. The Higgs mechanism was the way all particles acquired mass. Higgs proposed that the *Higgs Field* is coterminous with the universe. As particles move through the field, they cause discrete disturbances; the disturbances represent mass that the particles acquire. The Higgs particle (boson) is a localized excitation of the Higgs field, just at the photon is a localized excitation of the electromagnetic field. As particles move through the Higgs field, they accumulate Higgs bosons that cause a drag that becomes inertial mass. Hence, the Higgs particle, a characteristic of space, creates mass. With that observation science seems to imply that matter emerges from space. The Higgs field is a plausible description of reality, but it is not an explanation. My explanation follows.

G. Spatial Matter—My View <10>

The typical atom is 50,000 times the size of the nucleus and since electrons are *point particles*, the atom is mostly space and since protons and neutrons consist of quarks that are also point particles, we can say that matter is nothing but space. This implies that all that exists outside our minds (objective reality) is nothing but point particles and photons which in my view are nothing more than points of discrete space. The view that physical space is discrete makes it easier to visualize how matter could evolve from space. The idea of matter as a distortion of space is not a new idea. It is implicit, for example in: (i) super-string theory, (ii) quantum loop physics; (iii) the concept of the false vacuum, (iv) the theory that matter is created in the space warp near a black hole, and (v) Higgs mechanism theory. And what is a super-string but a gnarl in space? The hypothesis is that in some way a specifically structured segment of space forms a loop, vibrates with a characteristic frequency, and represents a specific particle. Such a scenario is easier to explain if physical space is discrete rather than continuous. I present a view of material reality as how it might exist at the explicative level. I begin with the *s-particle*.

H. The S-Particle

The basic particle of matter, the *s-particle,* is an infinitesimal volume of the cosmic s-frame in which the s-points organize much like a hydrogen atom that consists of proton and a single electron. Consequently, the s-particle is a fixed number of s-points gathered at the center of a volume devoid of s-points. The hydrogen atom is 30,000 times larger than the proton which means the hydrogen atom is mostly empty space and yet water that is mostly hydrogen is incompressible, implying that "empty space" creates materiality. The foundational structure of the universe that exists at the explicative level consists of vast but finite lattice of *s-points* (specific points associated with the rational numbers) which at each "instant of time" forms a fixed configuration, the cosmic s-frame. Throughout most of the universe the s-points exist in a uniform lattice of uniform spacing. Matter appears only in positions where emptiness appears in the otherwise uniformity of the lattice. And just as the emptiness between the electron and the proton in the hydrogen atom results in materiality, so too does the emptiness at locations in the cosmic s-frame.

Science's hope for the future is the "superstring theory." Implicit in that theory is the idea that strings are infinitesimal gnarls of space. Particles—both those constituting matter and those that carry a force—consist of various vibrating modes of the strings, and since strings create matter and energy, and strings are spatial, science has already arrived at the idea that matter is primarily spatial.

Chapter 6—Energy

Energy is the ultimate abstraction. Energy not only generalizes many observations, but the generalizations are factors in most physical phenomena. Energy is more mysterious than matter. We know how to describe it, how to use it, and how to measure it in all its forms, but we still do not know how to explain what it is. Mostly we think of it as a kind of substance pervading the entire universe that somehow causes things to move. We know that most of the energy available to us originated in the sun. Energy comes in many packages, and because of its transformability, energy is conserved, meaning it can neither be created nor destroyed.

For example, deep within the sun, when the fusion of hydrogen produces helium, and releases *nuclear energy* that bursts to the surface and becomes sunlight. After an 8-minute journey of 93 million miles, the radiant energy strikes the surface of the sea. The sunlight heats the molecules of water, vaporizes them, and lifts them to join countless other molecules of water vapor to form a cloud. After a journey across hundreds of miles, the water vapor in the cloud loses thermal energy, condenses into drops, loses a battle with gravitational energy, and falls into Lake Erie. Still in the tenacious grasp of gravity, the drops of water join a multitude of other water molecules and enter the Niagara River. About a mile upstream from the great falls, the water enters an intake and gushes through a cavernous aqueduct with increasing velocity until it crashes into the steel blades of a giant turbine generator causing the blades to spin in a blinding whirl of rotational mechanical energy. Connected to and spinning with the turbine are copper coils that plow through the magnetic flux of the generator's magnets. The lines of magnetic energy rip electrically charged electrons from the lattice of the metal, produce the electric current, and sends it on a final journey. The electrons flow across hundreds of miles of transmission wire, to reach the terminus of their journey in a tungsten filament. The electrons streaming through the lattice of tungsten excite the atoms to emit photons of radiant energy, the same form of energy as began its journey from the sun as light.

This sequence of energy transformations turns a small part of our private world from night into day. Scientists can precisely describe each of the transformations above with mathematical exactness, with which they can determine the nature and quantity of each form of energy involved; however, despite the accurate measurement of the various transformations, we cannot explain what energy is. Science tells us that the concept of energy is associated mathematically with other abstractions like force, mass, momentum, and work. In my view, I have no need to explore the various theories describing energy. My goal is to consider the main philosophical view of energy, namely. that energy: (i) appears in multiple modalities, (ii) conservation; and (ii) has equivalence with mass.

A. Forms of Energy

Energy grew out of the study of motion. Some scientists were interested in motion during the 17th century. Descartes, Newton, and Leibniz are the more famous among them. Descartes used "quantity of motion" to describe the nature of material motion; he found it was conserved. Newton found that quantity of motion combined with velocity that he called momentum was conserved. Gottfried Leibniz noticed that the sum of the mass of several bodies multiplied by their velocity squared was conserved. He called this quantity *vis viva* (life force) of the system. When later, in many cases, *vis viva* was not conserved, he realized that heat was lost when *vis viva* was not conserved; the heat that was generated by the mechanical motion. Eventually, heat was discovered as another form of *vis viva* and was not lost which led to the concept of energy, its many forms, and its conservation.

At the descriptive level of reality, science recognizes many forms of energy as described above in the sequence that began with nuclear energy in the sun and proceeded through transformations of energy from nuclear to radiant (sun light) to heat to gravitational to mechanical to magnetic to electrical back to heat and finally back to its original form in the light (radiant energy) from a tungsten filament. This differentiation of energy is possible not because of the scientific descriptions but because of the nature of the underlying reality.

Energy has served science well in its description of the dynamics of reality at the descriptive level. However, when examined more closely, we can view the various forms of energy in a more unifying way. We put the forms of energy into three general categories: ponderable, radiant, and potential.

Ponderable energy is mass in motion. The most usual form is kinetic energy, the product of mass and velocity squared. If the mass maintains a constant velocity, it neither gains nor loses energy. Mass loses energy only when it slows down, and because mass has inertia it is slowed only through the action of a force. Similarly, as Newtons' second law showed, a mass gains ponderable energy only when acted upon by a force, as though the mass absorbs the force and transforms it into ponderable energy. Heat, sound, and electricity are also forms of ponderable energy caused by the motion of molecules, air, and electrons.

Potential energy is the increase in the energy a body of matter assumes. The potential energy of a body is a measure of its spatial position in a force-field, or the increase associated with the configuration of its internal elements. For example, potential energy is created: (i) as internally stored gravitational energy when a bowling ball is raised to a higher position with respect to the surface of the earth; (ii) by increasing the distance between electrical charges; (iii) in a metal spring by increasing the tension (force) or compression (force) upon it. In addition, chemical and nuclear energy are potential energies. Chemical energy, stored in the chemical bonds (forces), is released, or gained during chemical reactions. The rearrangement of atoms and molecules due to the breaking or reforming of bonds causes chemical reactions along with a release of energy. Nuclear energy, stored in the nucleus of elements, is released in fission and fusion reactions.

Radiant energy has a dual nature; it is either an electromagnetic wave or a photon. A vibrating electromagnetic field can act like a particle. The vibrating electromagnetic field is composed of an electric and a magnetic component acting perpendicular to one another.

Energy is always part of every dynamic situation, and its ubiquitous nature leads science to conclude that it is the cause of dynamic situations when, in fact, energy is merely a manifestation of change (motion) at the descriptive level where science observes and models it. Energy is not the cause of the dynamics of reality, the cause exists at a deeper level, at the explicative level. The contention that If energy emerged from the singularity, it could not be the cause of the big bang. Something that existed in the pre-universe caused the big bang; hence, energy is an effect not a cause, and whatever caused the big bang must also be the source of energy.

B. Conservation of Energy

"That energy can neither be created nor destroyed" is one of the fundamental principles of physics. The woman who explained the principle, the German mathematician *Emmy Noether* (1882-1935), is among the unsung women ignored by the members of the scientific community. Her first theorem is stated in different ways, for example: (i) every differentiable symmetry of the action of a physical system has a corresponding conservation law, or (ii) in general any symmetry of a Lagrangian corresponds to a conserved quantity, or (iii) an element is conserved if it is symmetric under translation. No matter how phrased, Noether's theorem means: if an equation (*Lagrangian*) that describes an action does not change with a continuous transformation (is symmetrical), an action is conserved. For example: (i) if the transformation is a translation, momentum is conserved; (ii) if the transformation is rotational, angular momentum is conserved; and (iii) if the transformation is time, energy is conserved. Hence, when we refer to the conservation of energy, we must include the interchangeability of mass and energy because it is mass-energy that is conserved. The transformation of mass into energy and energy into mass is described mathematically but has not been explained philosophically.

C. Incremental Energy—My View

The static elements of reality—space and matter—are different at the explicative level; space is discrete not continuous, and matter is a configuration of discrete space, not an arrangement of solid particles of stuff. The dynamic elements—time, and energy—are also substantially different. Time and energy are abstractions at the descriptive level that originate at the explicative level as the incrementation of the cosmic s-frame. Time does not flow; time is simply the manifestation of incremental change in the cosmic s-frame. Energy is not the cause of motion; energy is a manifestation of incremental change in the cosmic s-frame. The incrementation is the result of a mechanism operating at the explicative level; I called it the holonomic mechanism. The holonomic mechanism is based on cellular elements, information, and an impetus of incrementation. Discrete space provides the cellular element; algorithms provide information, and God, not energy, provides the impetus.

Chapter 7—Time

The classical view of time belonged to astronomy. We define classical time by standards set by the motion of the earth and the moon. The revolution of the earth about the sun defines the year. The revolution of the moon about the earth defines the month. The rotation of the earth about its axis defines the day. The other measures—the second, minute, hour, and week—are related to arbitrarily chosen standards.

Newton said, *"Absolute, true and mathematical time which of itself, and from its own nature, flows equably without relation to anything external."* For Newton, time was an independent quantity that defined past, present, and future as clearly delineated domains that flowed at a constant rate. He believed that the equations of motion could be written based on the uniform rate at which the position of objects changed.

The classical view of time was simple and straightforward. Time was empty; it was the same everywhere. Time was infinite; there always was a past and there always will be a future; time ran on forever. Time was independent of the movement of matter; if the movement stopped, time would continue. Time did not interact with what it contained. Since events occur while time flowed independently, events could happen simultaneously. Time was continuous; it flowed smoothly; it was infinitely divisible. We can measure time by comparison to the regular motion of the earth, a clock, and the movement of a photon from an atom. Despite all of that, we must still ask: what is the nature of time?

A. The Nature of Time

Many philosophers addressed the question of time and broke into two camps: those that think time exists and those that argue that time does not exist. The "time does not exist" side of the issue includes philosophers such as the German philosopher Immanuel Kant (1724–1804) and others that took issue with Newton's absolute time. English metaphysician, John McTaggart (1866–1925), at the end of the 19th century and the beginning of the 20th century authored a book called the *Unreality of Time*. He postulated events arranged in two ways that he called the A-series and the B-series of time. The A-series is a sequence past, present, and future while the B-series is simply "earlier-than" and "later-than." The A-series models time as a dynamic flow. In the fixed B-series, an event that happens before another event will always remain in that relationship. Washington will always be president earlier than Lincoln. McTaggart's argument was that the A-series generated the B-series and therefore time did not exist because the A-series was inconsistent in that every event is present, past, and future depending on the point of view.

When a person smiles at you, the event when experienced by you, i.e., in your present, is the other person's past because the smile happened before you experienced it. The tense view of time, past, present, future represents the *flow* of time, and it is the flow of time, not time *per se,* that causes the dilemma. The flow of time as described by the sequence past, present and future implies that the future becomes the present and the present becomes the past. However, when time is continuous, it is infinitely divisible and that implies that the intervals become infinitely smaller until the interval that represents the present disappears into a dimensionless line between past and future, hence the present does not exist.

Only the transition from past to future occurs. Still, we mentally perceive the present to possess some duration. The elimination of the present has more to do with the assumption of continuity than does our experience of reality. Continuity reduces the present to a point in the flow of time. Because we perceive the present as an interval of time, the present must naturally exist as an interval of finite duration as it would if we assume time, like space, is discrete. The flow of time problem arises from the assumption that space and time are continuous. Although we may conclude that time does not flow, we still conclude that time is real. Time is the perception of change; time does not exist unless there is change and furthermore the change must be periodic, i.e., recurring uniformly. Periodic change allows the construction of clocks.

Clocks are devices or natural observations that possess observable uniform and periodic change. Were there not uniformly periodic change in nature there would be no clocks and time as an element of reality would not exist. If the rotation of the earth were random, we would still have the periodic change we call the day, but it would not be uniformly periodic change and the randomness of the length of the day along with the increments of hours could not allow the measurement of time. It is difficult to imagine how reality would exist if God did not make it periodic.

B. Space-Time

The classical view was simple and intuitive and then along came Einstein. Einstein assumed the constancy of the speed of light and that the laws of physics are covariant and using those assumptions, developed the special theory of relativity from which he merged space and time in a space-time continuum. This theory led to the shift from the classical to the modern paradigm and introduced previously unimaginable ideas such as the shortening of length and slowing of time with increasing linear velocity. Most amazing was the idea that time could flow at different rates and there was no such thing as simultaneity. Scientists have experimentally demonstrated the slowing of time many times. Therefore, Einstein was right; however, he was right only so much as we stick with the descriptive level of reality and not ask *why* things are the way they are? I ask that in Einstein's contention that time slows as linear velocity increases; might it not be that it is the clocks that are slowing, not time itself?

C. Incremental Time <11>

I contend that objective reality consists of an incrementing cosmic s-frame, a 3-dimensional configuration of s-points in a relationship to one another. Within the cosmic s-frame, matter consists of certain local configurations of s-points that expose emptiness (volume with no s-points). The s-points exterior to those local configurations form a uniform 3-dimensional lattice that we experience as universal space. When a sequence of symbols {N-n} that represent an increment occur, energy and time are created at the explicative level, time is incremental as measured by certain configurations of s-points we call clocks. The time we experience is *cosmological time.* The smallest increment {N-n} is equal to ~10^{-43} cosmological seconds; it is known as the Planck era and in one second, 10^{43} increments occur.

Cosmological time is the manifestation of change; a cosmic s-frame represents a stationary universe in which nothing appears to change and for one Planck era, reality is associated with the relative change of position of s-points from one s-frame to the next s-frame within the cosmic s-frame as measured relative to cyclical changes of s-points configured as clocks.

Cosmological time, the time we experience, began with the big bang and advances in finite increments we call intervals. Materialists argue that time emerged from the singularity with the big bang, that there was no "before" because there was no cosmological time before the big bang hence there was no beginning and no need for God. In a holonomic universe, I argue that the infinite nothingness that always existed is the Mind of God. God exists; an irrefutable premise allows for a second kind of time that I call "ontological time."

Ontological time is associated with what happens in the interval {N-n}. Ontological time is immeasurable because our clocks do not 'move' during {N-n} and hence cosmological time does not advance. However, at the explicative level, the imperceptible change from {N} to {n} must occur in the Mind of God that represents a second modality of time that I call ontological time.

God's mind (infinite nothingness) is not and cannot be empty. In Part IV, I will argue the concept of "Imago Dei," made in the image of God, that we have a psychical element, a soul, made of psychical stuff. We know that the psychical stuff – *nous* - that forms our mind, contains qualia consisting of feelings and sensations, and intellect containing concepts and percepts.

A train of thought in our mind proceeds sequentially with one thought following another. Spontaneous thought changes the train of thought and induces a new train of thought. Because it is certain that God thinks. thoughts represent a form of change that can define a second form of time, hence ontological time.

Ontological time is continuous and like McTaggart's A-time in which past-present-future flow without stopping, the present does not exist. Cosmological time is like B-time; events establish a before/after relationship and there is a "present" that exists within the {N-n} interval, currently equal to a Planck era. There is a flow of ontological time in which the present disappears, which is another way of saying that ontological time does not stop because in addition to pure existence, God unlike our lesser existence that is a sequence of incrementations of static {N}'s, God is absolute being, a continuous flow of the psychical. God's Mind that has existed eternally, is the clock that measures ontological time. Since our minds perceive only cosmological time, it is conceivable that the duration of ontological time that fills the {N-n} interval is different from the interval of cosmological time that clocks measure. We experience a duration of time because of the psychical part of our mind. There is no way to determine the duration of ontological time within each interval of cosmological time; it could be much longer than the Planck time and it might even vary without effecting the uniformity of cosmological time.

Chapter 8—Consequences

Postulating God's existence allows me to imagine how He might have created and sustains reality with what I call the model of dual reality (MDR). The MDR: (i) is based on a foundation of discrete space instead of continuous space; (ii) is based on change being the result of information rather than energy; and (iii) is based on algorithms instead of mathematical equations. The MDR is a 3-dimensional form of a cellular automaton in that

the lattice like arrangement of s-points in the cosmic s-frame functions as cells; the information and algorithms provide the self-activating nature of its dynamics. As a result, configurations of s-points form everything—including living matter—with a single mechanism, and the holonomic mechanism, operating at the explicative level, generates all phenomena, including life. I do not claim that the MDR is an explanation of reality, but I offer it to serve as a guide for how we might find a final explanation. The premise, God exists, allows us to explore possibilities that secular humanists, who are restricted by the paradigms of science, are unable to consider. It also allows answers to certain questions that modern science cannot, does not, or avoids answering. I now apply my view at the explicative level to answer questions associated with science's view at the descriptive level of reality

A. Questions

Science is an endeavor in which a description of an observation suffices; explanation is not required. For example, the interchange of matter and energy is described mathematically and demonstrated experimentally, but there has been no explanation of how the interchange occurs. Is there a mechanism? On the other hand, what is the underlying explanation of the *wave-particle duality*? Or what is the nature of a *field* that forms the substrate for much of what science observes? Or what is the *quantum field* from which virtual particles pop in and out of existence? Science has not addressed such questions because paradigmatic science does not allow for the possibility of a psychical substrate. On the other hand, the MDR does allow explanations of such observations. The success that science has had in describing the microcosmic has been magnificent. However, several inexplicable observations have appeared, such as: (i) the Big Bang and the expansion of space, (ii) the constancy of the speed of light, (iii) the source of the laws of physics, (iv) wave-particle duality, (v) *non-locality*, (vi) *entanglement*, (vii) the double split experiment, and (viii) Schrödinger cat experiment.

1. The Laws of Physics

One of the questions that philosophers sometimes ask is: what is the source of the laws of physics and why is reality so coherent that it can be described by mathematical equations? For example, why is gravity described with the inverse square law? Science or philosophy has no answer. But the model of dual reality provides an explanation. The explanation derives from the assumption that universal space is discrete. Discrete space is associated with the rational numbers that are based on the integers and the ratio of integers. In the model of dual reality, the cosmic algorithm explains force, mass, charge, inertia, and many of the concepts of physical science. Likewise, the elements of the cosmic algorithm are the source of the mathematical equations of physics. This answers the question as to where the equations exist; they derive from the cosmic algorithm that is in the noumena in the realm of potentiality.

2. Expansion of Space

According to the Big Bang theory, the space that creates the dimensions of the universe emerged from the singularity. The common view implies that the stars and galaxies are not moving through space, but space is being added, and space is pushing the galaxies apart. This cannot be because: the only possibility of expanding continuous space that is already infinitely divisible is by adding more space. So where would the additional space come from? I contend that discrete space offers a more plausible and easily imagined explanation of the expansion of space because expanding the s-gap expands universal space. Also, moving a point in space does not require energy; the expansion of the s-gaps is the result of a cosmic algorithm and an impetus to increment in the Mind of God.

3. The Singularity

If matter is the absence of discrete space, then the singularity represents an enormous cosmic s-particle the mass of which is determined by the number of s-points in the singularity. What that implies is that the singularity contained the entire universe, and energy and time emerged with the first incrementation. No energy is required to move discrete space. No quantum fluctuations or negative gravity is needed. And the singularity can only be explained as a sufficiently large collection of dimensionless s-points squeezed into an infinitesimal volume.

4. Gravity

When the cosmic algorithm produces two fixed number of s-points configured as matter (mass) to reduce their separation a fixed number of s-gaps in an affixed number of increments (time), what is manifested in the common view is as a motion of matter caused by a "force" called gravity. In my view it is the cosmic algorithm supplying "information" applied to the divine impetus. Other forces such as those associated with properties such as electric charge and magnetism can be explained similarly.

5. Space-Time

I have run across this question: "How can space and time be the same thing? My answer is that when the cosmic s-frame increments, both the s-points that define space (and matter) move simultaneously with the clock s-frames that define time, hence time is nothing more than the reconfiguration of discrete space as measured on simultaneously incrementing clocks.

6. Matter—Energy Interchange

The cosmic s-frame is the substrate for localized configurations of s-points that form matter positioned amid the undisturbed crystalline lattice of s-points that represents *open-space*, the space between bodies of matter. At the explicative level, s-points are the ground of both matter and space. Kinetic energy is the repositioning of the configuration of s-points that forms matter. Radiant energy occurs with incrementation of s-points that are associated with open space. Since both matter and energy are manifestations of the incrementation of s-points, the mechanism that explains the conversion of matter into energy and energy to matter is simply based on the reconfiguration of s-points. A local configuration representing matter can be reconfigured as a local configuration representing energy and vice versa.

7. Conservation of Matter–Energy

The singularity contains a finite number of s-points from which both mass and energy are formed. There is a finite number of s-points designated for the creation of mass-energy, so when one is increased the other decreases.

8. Wave–Particle Duality

Bohr's complementary principle explains the duality of light as being dependent on the manner of observation, but how does observation determine whether light is a wave or a particle? Imagine light as a wave-like configuration of s-points for part of a sequence of increments and as a particle-like configuration for a subsequent interval. (What those configurations look like I am leaving for future generations of physicists to workout). Since the cosmic s-frame increments 10^{43} times every second and the frequency of gamma rays on the high end of the frequency spectrum is in the range of 10^{11} cycles per second there is time to fit the intervals taken up by the particle like configurations. In other words, a light beam consists of an incrementing sequence of alternating waves and particles. Waves and particles both exist physically at different times on the path of incrementation. Moreover, the method used to observe selects the appropriate particle or wave configuration. This interchange of particle and wave configurations with each {N-n}, explains the wave-particle duality.

9. The Constancy of the Speed of Light

One of the great constants is the speed of light. Why is it constant? However, given an impetus, a static cosmic s-frame, {N}, increments and becomes a new configuration of s-points, {n}, and {N-n} represents the minimum time interval of incrementation, equal to 10^{-43} seconds, the so-called Planck era. The configuration of s-points that form a photon changes position every single incrementation {N-n} and when combined with the approximate minimum average displacement of s-points of $\sim 10^{-35}$ meters, a constant speed of light is determined. I assume that the cosmic s-frame increments 10^{43} times per chronological second and the s-gap varies over time, starting at ~ 0 in the singularity and increasing uniformly in time to the present value of the Planck length $\sim 10^{-35}$ meters. The maximum displacement of s-points from one s-frame to the next is the Planck length; the velocity of apparent motion has an upper limit, hence the constancy of the speed of light. Since the speed of light is dependent on the spatial distance of the s-gap, then the velocity of light increases as the s-gap widens with the expansion of the universe. The speed of light was zero at the first instant and had been increasing ever since. This might explain the observation of the apparent acceleration of the expansion of the universe.

In addition, the apparent motion of objects—photons or s-particles—is determined by the incrementation rate, the number of times the object is reconfigured. The incrementation rate for photons is equal to the inverse of the Planck era, $\sim 10^{43}$ times in a cosmological second. The photon is displaced with a Planck length, $\sim 10^{-35}$ meters, with each incrementation. S-particles, being a component of larger matter particles, are displaced the same Planck length each time it is reconfigured but the incrementation time is longer (increments less often) and the number of incrementations per second and hence the speed is lower.

10. Curvature of Space

How is it possible to curve continuous space? Mathematicians have developed methods for defining the curvature of space mathematically, and Einstein used Riemann's method to describe the curvature of space in proximity to large masses. However, the equations do not curve space; curvature exists because of the influence of mass. Why the curvature of space in the proximity of large masses remains a question not answered or even asked. The mathematical curvature of space is understandable; physical curvature of continuous space is not. In flat space, far from the influence of proximate mass, trajectories follow straight lines. In the proximity of a large mass, the straight-line trajectory still exists in continuous space, but a particle moving in that proximity would follow a curved trajectory. Why would it not take the straight trajectory? The Newton explanation was the presence of a force. Einstein claimed that a particle would follow the curvature of space despite the availability of a more natural straight trajectory. The curvature of space in the proximity of large masses describes the observations; but space is not curved, it is trajectory of light, an immaterial phenomenon, on which the force of gravity should have no effect. But there is no way to explain the curvature of an infinitely dense substance like space.

My view: The only answer to this dilemma then is that—as Zeno discovered eons ago—universal space cannot be continuous; it is far easier to imagine Einstein's proposition that physical space bends in the proximity of gravitational masses by imagining a deformation of s-gaps in discrete space. Even though the assumption of continuous space describes the curvature of space mathematically, it does not explain it physically; only the formability of discrete space explains the possibility of curving physical space.

11. Dark Matter—Dark Energy

The edges of galaxies have velocities greater than the capability of the gravity of the observed mass to hold the galaxy together. Also, the light to mass ratio of galaxies is greater than the same ratio of the sun, the standard. Cosmologists resolved this anomalous observation by abstracting the existence of dark matter. Dark matter accounts for about a quarter of the matter needed to explain observations. But what is Dark Matter? Science has no firm description of its nature, but the leading proposal is a weakly interacting massive particle (WIMP) that interacts with gravity.

At the descriptive level, to satisfy the requirement for spatial flatness, when calculated with the relativity equation, the curvature of space must be zero. The flatness requirement places stringent condition of the mass-energy tensor. There is not enough luminous mass to produce the zero curvature to satisfy the equation, so scientists abstracted dark matter and dark energy to supply the required missing matter-energy. The result is that only ~5% of the total mass of the universe is luminous; ~ 25% is dark matter and ~70% is dark energy. These abstractions work very well in describing observations, and science is perfectly satisfied with describing their observations without explaining why things happen the way they do. The big question is what is dark matter and what is dark energy? Modern science has no answer.

My view: **dark matter** is simply the manifestation of s-particles. They are smaller than an electron, so they are spread throughout the universe (the cosmic s-frame) and are not seen. My hypothesis is the s-particle is mass, an infinitely small mass. However, there could be s-particles of various sizes. **Dark energy** is the vibrating motion of the s-points that form universal space, i.e., s-points not engaged in the formation of s-particles (matter).

12. Fields

Michael Faraday's postulation of the field concept replaced the luminiferous other as the explanation used during the classical period to explain the propagation of light as a wave. In addition to the question, what waves? One must wonder, what is the underlying nature of space that allows an electromagnetic field to exist in the same space as the gravitational field? Science feels no need to "explain" why things happen because scientists are satisfied merely to "describe" how things happen, especially when observation confirms that it does happen.

My view: the field concept is verification that space is discrete because as a 3D lattice of s-points it is the only way that cosmic s-frame can be reconfigured, and change introduced. Also, it is prolific with properties that are not explicable, only describable. If one believes God exists, it is possible to understand how physical fields that are prolific with electro-magnetic, gravitational, quantum, and Higgs fields, all reside in the same space without interference. The arrangement and reconfiguration of the s-points determine the nature of the field. Because the extremely small spacing allows different fields to co-exist.

13. Higgs Field

Several physicists aided in developing the Higgs Field to explain the mass of the W and Z gauge bosons that mediate the weak force. Science observed and verified the existence of the Higgs boson. There are no explanations for the field concept of physics, but it is useful for creating mathematical descriptions. Why all particles acquire mass when moving through the Higgs field is not known?

My view: discrete space forms the cosmic s-frame (the universe). The formation of matter by the reconfiguration of s-points allows a particle of matter to accumulate more s-points at the center of an s-particle as it changes position incrementally.

14. Entanglement

Entanglement is a result of the *Pauli exclusion principle* that states two or more fermions with same quantum numbers cannot occupy the same space, so, one particle determines the nature (quantum numbers) of a second particle. For example, spin is a quantum mechanical property discovered in 1925 to explain the *Zeeman Effect*—the splitting of spectral lines for atoms placed in a strong magnetic field into two closely separated fine lines. Electrons have spin values of $-\frac{1}{2}$ or $+\frac{1}{2}$. If we separate two electrons described by a single wave function, no matter how far—even to opposite sides of the universe—according to quantum mechanics, they continue to remain a unified system described by a single wave function. Then, if the spin for one of the electrons changes to from $-\frac{1}{2}$ to $+\frac{1}{2}$, the spin of its partner—according to the exclusion principle—simultaneously changes from $+\frac{1}{2}$ to $-\frac{1}{2}$. This result of quantum mechanics is contrary to that of the concept of locality on which classical physics—including relativity—is based. My view: each type of particle corresponds to a specific and uniform configuration of s-points i.e., the s-points of all electrons are configured alike, but are spatially unique, i.e., they cannot reside in the same space (the exclusion principle) because each s-point is referenced to a single rational number. This explains the why the exclusion principle works, how the information of the change in spin arrives at distances greater than the distance life travels is explained in the non-locality section that follows.

15. Non-Locality

In 1935, Einstein with the Russian physicist *Boris Podolsky* (1896–1966), and the Israeli physicist *Nathan Rosen* (1909–1995) presented a paper known as the EPR paradox. As a determinist, Einstein was critical of the quantum mechanical view. He conceived the EPR paradox to demonstrate that quantum mechanics violated the principle of local cause (locality) and was therefore an incomplete theory. Locality means that a cause-and-effect event can take place only for the mechanically connected events, i.e., the physical separation of the two events must be within a distance no farther than the distance light can travel, i.e., they are within a common light cone.

In 1957, the Irish physicist, *J.T. Bell* (1876–1963) devised an experiment for settling the argument. His experiment involved polarized light (photons) split into two parts after the manner of an EPR experiment. Two Americans, *John Clauser* (1942–) and *Stuart Freedman* (1944–2012) performed the experiment in 1972. They compared the measurements to statistics calculated using quantum mechanics. The observations were in perfect agreement with the calculations leading to what is known as the Bell theorem, which states that if the statistical predictions of quantum mechanics are valid then the principle of locality cannot be true. Given the verification of quantum mechanics statistics by a great variety of phenomena that were 100 percent correct, the only logical conclusion is we live in a world in which non-locality exists and information can travel faster than the speed of light.

My view: The Planck distance divided by Plank era determines the speed of light. However, at the explicative level of reality, information can travel through the continuous pneuma of the realm of potentiality where the Plank era is infinitely small so the speed of information transfer is infinite, and information can travel faster than light and hence the explanation of non-locality and entanglement.

16. Schrödinger's Cat Experiment

Schrödinger's view of matter waves as probabilities led him to conceptualize his famous cat experiment. Consider a cat in a box containing a source of radioactive decay next to a cyanide pellet. If the device randomly emits a photon that strikes a mechanism releasing the cyanide. According to the common view, the cat is either alive or dead whether an observer opens the box or not. In the *Copenhagen Interpretation*, the cat is neither dead nor alive until the observer opens the box. In the *Many–Worlds Interpretation*, the world in which the cat is alive exists along with a world in which the cat is dead. Hence all quantum outcomes exist separately in parallel worlds. The implication is that the cat's fate depends on the action of an observer, a phenomenon that has no scientific explanation, but it does have an interesting implication, namely, that a psychical event (observation) can affect the material world (the death of a cat).

17. Materialism vs. Idealism

My view of reality resolves the contrary views of materialism and idealism because both are explained based on a single principle, space. And nothing is more basic than space unless it is numbers from which the spatial dichotomy derives. One can imagine that the discrete space that is derived from the rational numbers is the materialistic principle while the continuous space that is derived from the real numbers is the idealistic (immaterial) principle.

PART 2—Subjective Reality

Chapter 9—Totality

I shift the theme in Part II to the "nature of becoming," as I apply the model of dual reality to the historical events of the cosmos, as it expanded from the singularity to the formation of the universe at which point the material world diverged along a new path of complexity that culminated in the creation of subjective reality.

The goal is to provide background for the appearance of subjective and eventually rational and transcendent reality, and to further develop an argument with which believers can refute the materialists' implication that God does not exist. Unlike objective reality that encompasses the entire universe, subjective reality appears in a miniscule part of the universe—in sentient animals, discrete organisms with the amazing property of consciousness. Subjective reality did not exist until the *pneuma* was *corpuspirated* to become *nous* in a multicellular organism around 540 mya.

> Note 18: I coined the word corpuspirate to mean "to enclose an immaterial substance (pneuma) that is called nous within a multicellular body."

The evolution of universe that began with the singularity represents the sequential instantiation and actualization of possibilities that tracks increasing material complexity. The chart depicts the complexity of matter plotted against the history of the universe the goal of which is the creation of the human mind. Science is stuck with the abstraction "it emerges from the brain." In my view, I go a step further and use the stuff the ancient Greeks called "nous" to explain sentience.

> Note 19: Sentience is the experience of sensations such as sight, sound, smell, taste, touch, feeling, and emotion. Explaining and even describing sentience is impossible, it can only be experienced.

Having described the constituent physical parts (space, matter, energy, and time), I now put the parts together as my view of Totality. Science recognizes a single realm of actuality consisting solely of the physical universe which despite an absence of a psychical component, includes—as discussed in Chapter 2—objective, subjective, rational, and transcendental realities. In Chapter 6, I introduced the cosmic s-frame as consisting of two realms, actuality (the universe) and potentiality (pneuma). I introduced realm of possibility in Chapters 1 and 2 as the pre-universe, the before/beyond, or the infinite nothingness that is the Mind of God, and the source of qualia, intellect, and all possibilities. Here I deal with how God created them.

Before a thing was actual, it was potential. Before a thing is potential it was possible, another realm. Hence, instead of one realm of actuality, my view expands to include a psychical component that is the basis for the realms of possibility and potentiality. Science, in ignoring the existence of a psychical component, has had little success in explaining psychical phenomena such as consciousness, mind, thought, and qualia, the essential elements of subjective and rational reality. Based on the premise of the existence of God, and the consequent psychical element called infinite nothingness, my view of Totality is as follows:

From the eternal past until 13.7 billion years ago (bya), prior to the big bang, the pre-universe, consisted of pure existence called infinite nothingness and was, as far as we can know, a single realm. Then an incredible event took place when an infinitesimal object called the singularity rapidly expanded to form the universe in an event called the big bang, and reality divided into two realms: the universe, and the pre-universe. The universe consists of discrete space that creates matter and defines its spatial dimension.

A. The Realm of Possibility—Mind of God

As shown in figure 1, the cosmic s-frame consists of the universe plus the coterminous pneuma. Hence, the cosmic s-frame consists of two realms: the universe is the realm of actuality, and the pneuma is the realm of potentiality. The universe is the physical realm, the pneuma is an immaterial or psychical realm. Now Totality consists of three realms: possibility, potentiality, and actuality. As discussed in Chapter 2, I also viewed reality at four levels: objective, subjective, rational, and transcendent. Objective reality inheres in the realm of actuality; subjective reality mostly exists in the substance (nous) of *metazoa* including humans; rational and transcendent reality exists only in the human mind.

The *realm of possibility* (or just Possibility) necessarily includes all that is possible, and we can only know actualized possibilities, i.e., have become real. Only God knows the full set of possibilities. The realm of possibility existed before the realm of potentiality and the realm of actuality. It appears in this polemic in several disguises. The realm of possibility also: (i) can be referred to as: the pre-universe, or the before/beyond, (ii) is pure existence, absolute being, eternal, dimensionally infinite and infinitely divisible, (iii) made of continuous space; (iv) is perfect, formless, infinite, eternal, immutable; and as described later, omnipresent, omnipotent, and omniscient; and (v) stretches from the edge of the expanding universe to infinity; and (vi) is filled with infinite nothingness, the psychical substance from which we owe our existence. It came before and lies beyond the universe hence It is the *Mind of God*.

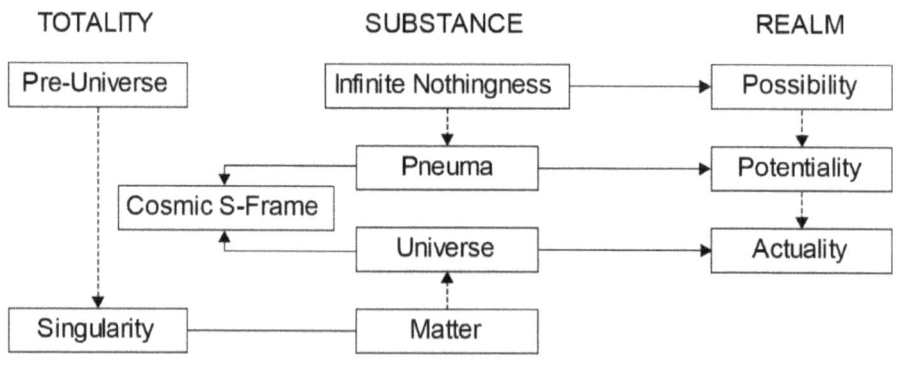

Figure 1. Structure of Totality

B. The Realm of Potentiality—Pneuma

God created the singularity, the object that consisted of all the s-points crammed into an infinitesimal volume that expanded to form the universe. As the universe expanded, it enveloped the infinite nothingness from which it emerged and formed a cosmic s-frame. The cosmic s-frame consists of the universe plus the infinite nothingness in which it exists. I refer to the infinite nothingness that is coterminous with the universe as *pneuma*; it is coterminous with the s-points that represent the material part of the cosmic s-frame. The pneuma forms a new realm called the realm of potentiality; it is distinct from the realm of possibility from which it formed, and contains those possibilities, called *noumena*, with which God creates and sustains the universe. It is through the actualization of noumena that God creates and sustains objective reality.

Noumena are psychical entities instantiated from the realm of possibility to reside in the realm of potentiality. Noumena are protypes of phenomena. There are four modalities of noumena: foundational (*f-noumena*), conceptual (*c-noumena*), perceptual (*p-noumena*), and transcendent (*t-noumena*). Holonomic noumena are the possibilities that are instantiated as the informational input to the cosmic algorithm that actualizes objective reality. I use the word *realization* to describe the transition from noumena to intellect; the conceptual noumena realized as concepts in the mind of humans, and perceptual noumena realized as percepts in the nous of metazoa. Noumena are what *Plato* (c.428–348 B.C.) meant by *ideas* or *forms*. What I am proposing is that pneuma, not consciousness, is the substance that is described as *panpsychism*.

Rational reality consists primarily of: (i) conceptual intellect; that is associated with digital concepts such as words, numbers, musical notes; (ii) forms that exist as neural maps in the brain; and (iii) associated meanings. Concepts are instantiated from the realm of possibilities and stored in the realm of potentiality as *c-noumena*. The timing of the discoveries (realization of noumena) depends on many factors that determine the probability of discovery. Probability determines the history of science and technology. However, once the initial discovery occurs, the probability of a repetition and discovery of that concept increases greatly. The enhanced probability of proliferation of the discovery is like a phenomenon called *morphic resonance,* an idea introduced by Ralph Sheldrake in his book of the same name.

Levels of reality occurred sequentially as the complexity of matter evolved. Objective reality has existed from the singularity to the present; subjective reality began with multicellular life; rational reality began with the human mind; and transcendent reality began with the soul. I discuss the connections among the realms from possibility to potentiality to reality in the next section. Possibilities, that are instantiated as noumena, function as the pattern or source for the actualization as phenomena. I use the term *instantiate* to mean brought into the realm of potentiality, and I use the term *actualize* to mean brought into reality. Unlike the scientific view of reality that is a bottom-up approach, my view is a top-down view that allows a noumenal pattern to guide the evolution of material complexity, a more direct approach that explains much of the formation of physical structures such as galaxies and the mechanisms associated with psychical phenomena such as *ontogeny*. It also allows a top-down approach for the experiences of qualia and intellect.

Science does not have to deal with the actuality of reality because in the common view, things just show up or simply come into being. The reality that science describes is matter in motion; that is due to energy; that creates motion; that appears as time. Science uses mathematical equations to describe the interaction of material particles occurring in a background of space. This works well because the interaction of matter as described by equations explains the buildup of material complexity that evolves as objective reality. It is though the universe created itself, and science is happy with that.

When life appeared, phenomena for which the equations do not apply, confronted science. Life expanded and psychical phenomena such as consciousness, mind, and thought emerged. Although science recognizes that subjective reality is psychical, science has not found plausible descriptions let alone explanations. Science views the dynamics of reality as the maelstrom of energy in which all movement is the result of the action of a variety of forces.

This thesis is based on the premise that God exists, and my approach is to describe how God creates and sustains reality. God took 13.7 billion years to build a foundation with which to create the human mind. I will describe how that occurred in the next several chapters. The story is described as a path of increasing material complexity. This raises the question as to how God communicates with His creation. I, of course do not know how, but because my thesis claims to describe how God sustains reality, I feel obliged to devise a plausible scheme that describes the exchange of information between God, the realms, and the levels of reality. I describe God's interaction with the universe as a set of four information channels, shown in figure 2. Information flows through each channel from the infinite nothingness, to the pneuma in the realm of potentiality. From there God uses the foundational channel to directly create and sustain the realm of actuality The figure 2 chart depicts an s-frame (our body) representing a material particle of objective reality, *corpuspirated* with nous. Although not shown, both the realm of potentiality and the realm of actuality occupy the same universal space. The universe is the material part and the realm of potentiality—the pneuma is the psychical part; together they form the cosmic s-frame that exists within the nothingness of the realm of possibility, the Mind of God.

> Note 20: I created the word corpuspirate. It means to fill with a psychical substance: (i) bios for cells, and (ii) nous for multicellular bodies.

Yes, God is omnipresent; each particle of matter is immersed in a cosmic ocean of psychical substance called pneuma. This is the source of panpsychism. There are four channels through which information flows. They are: (i) foundational; (ii) sentient; (iii) sapient; and (iv) transcendental. There are three realms: (i) possibility; (ii) potentiality; (iii) actuality, plus the human mind. There are four functions; (i) instantiation is the transfer of information from the realm of possibility (the Mind of God) to the pneuma to form noumena in the realm of potentiality; (ii) realization is both human *discernment* and learning that brings conceptual and perceptual noumenal to mind; (iii) actualization is the use of concepts and percepts, that are stored in the mind, to create ideas in the realm of potentiality and material forms in the realm of actuality; and (iv) resonance is the action in which the mind directly engages transcendent noumena to induce qualia in the form of emotional feelings. Humans, created in the image of God, have been given the power for creating reality through two information channels, a conceptual channel, and a perceptual channel.

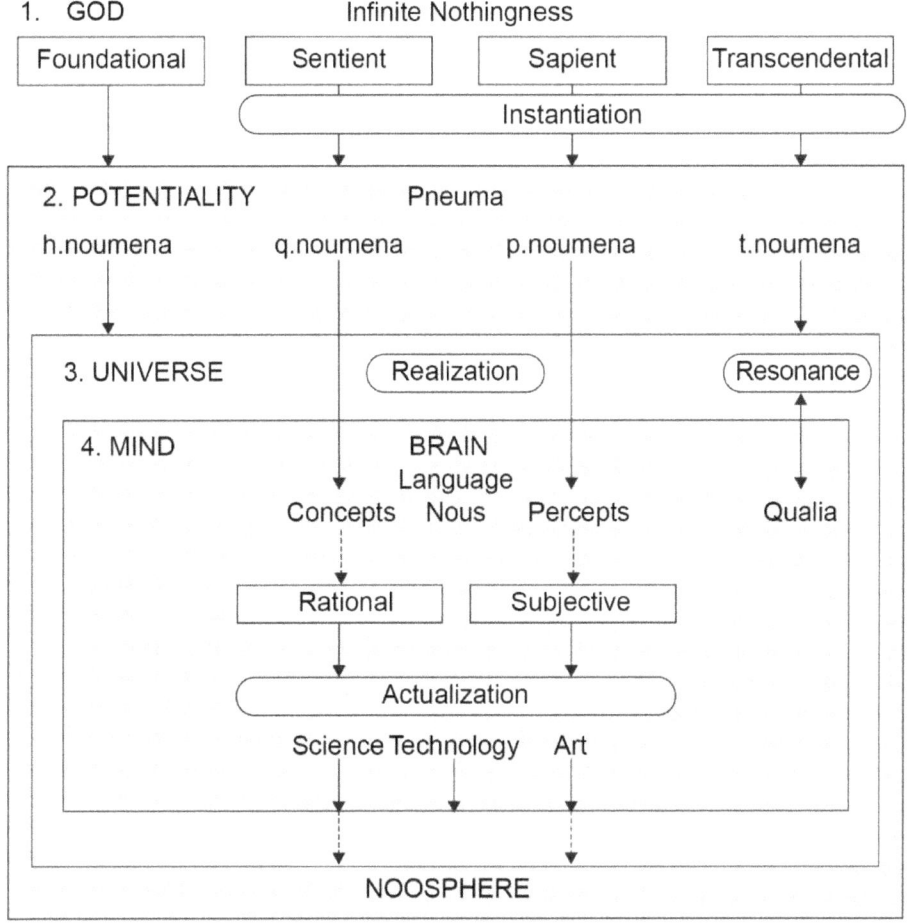

Figure 2. God's Information Channels

1. The Foundational Channel

God used the foundational channel—the only channel that existed for the first 12+ billion years—to create the universe. The possibilities associated with the creation of the universe were instantiated (downloaded) and stored as foundational noumena (f-noumena) in the pneuma of the realm of potentiality. F-noumena consist of a set of rational numbers (the input) and an algorithm that sequentially creates objective reality using the holonomic mechanism.

2. The Sentient Channel

God was directly only involved in creating and sustaining a continuous growth of objective reality through the foundational channel until sentience was created. By the term *sentience* I mean, "to be alive and capable of experiencing sensations." For eons before sentience emerged the universe was without light, sound, and all other sensations. At its beginning, sentience was barely noticeable and did not reach a substantial level until well after the appearance of multicellular life. The creation of the stars and galaxies occurred with enormous release of energy but there were no sentient beings to *see* the fireworks; there was no *light,* it was absolute darkness. Light was waiting for the creation of eyes to see.

The universe expanded for some time before the universe created the realm of potentiality. Discrete space enclosed infinite nothingness that became pneuma that then became the foundation of subjective reality. Amorphous (without form) possibilities that were always present in the infinite nothingness of the pre-universe were instantiated as perceptual noumena (p-noumena) residing in the pneuma to await the presence of *metazoan* nervous systems sufficiently cephalized (concentrated in the cranium) to actualize the *subjective experiences,* the elements that form subjective reality such as sensations and feelings. The common view does not recognize an *immaterial memory* in which the realization of p-noumena is stored.

3. The Sapient Channel

God created an information channel in which possibilities are instantiated to form conceptual noumena (c-noumena) that reside in the realm of potentiality. C-noumena are the discrete blueprints for the 4-physical elements and all other discrete relationships that science describes. The mind discerns the c-noumena and through realization transforms them into concepts. Words, the basic element of language, are also the basic requirement of thought. We can assume that the *Homo sapiens* that preceded the first human were, like all metazoa, sentient. Sentience means to be alive and capable of experiencing sensations. The first human was also sapient. Sapience means to be aware and able to think. To think requires a word, and words require a mind. Thinking or at least a mind arrived about 50-thousand years ago as the first *Human* emerged from the *Homo sapien idaltu.*

4. The Transcendental Channel

Humans are capable of directly experiencing transcendent noumena (t-noumena) such as beauty, happiness, truth, and justice in the mind; but also form transcendent reality as a common experience among multiple minds. We experience transcendent reality through wonder and its associated activities such as discernment and appreciation. Wonder can generate a direct resonance with t-noumena in the realm of potentiality. Simply observing or hearing a thing of beauty is not sufficient for a full experience. Seeing or listening may generate a pleasant experience, but to provoke the full experience of beauty, the active mental faculty of the mind must wonder and discern, and the mind resonates with a feeling of awe, the full experience of beauty.

Chapter 10—Reality Modes

The common view recognizes four levels of reality: objective, subjective, rational, and transcendental. Objective reality has existed from the big bang to the present; subjective reality began with multicellular life; rational reality began with the human mind; and transcendental reality began within the human soul.

A. Objective Reality

Objective reality is the physical part of the cosmic s-frame. Since objective reality forms from s-points of discrete space, it is a digital realm in contrast to the analog realms of possibility and potentiality.

Consider an apple sitting on a table: We see a smooth, round, red object, a smooth texture, a form, and a color. However, the apple consists of discrete elements such as carbon and oxygen; carbon and oxygen consist of molecules; molecules consist of atoms; atoms consist of protons, neutrons, and electrons; and protons and neutrons consist of quarks. We do not see that hierarchy of matter; we see a smooth three-dimensional red form that we recognize as an apple. It is not the light reflected from the skin molecules of the apple that we see; light is not red or any color. Light is the wave motion of a spatial field or the motion of particles of light called photons. Spatial fields and photons have no color. When light from the apple enters our eyes, the lens focuses the photons on the retina at the back of the eye where specific neurons are excited and send electric impulses along peripheral nerves to the various parts of the brain. In the brain when specific neural maps are exciting, we perceive an image of the smooth, three-dimensional, red form that we recognize as an apple. The question is: which is real, the configuration of quarks and electrons sitting on the table that we *see* or the image of a smooth 3-D red object that we perceive? Both exist and are therefore real. However, later I argue that objective reality exists as spatial configurations of quarks, electrons, and photons in motion and the perception of redness is another order of reality, subjective reality. Although existing in radically separate ways, there is an intimate connection between the two orders. The configuration of quarks, electrons, and photons that form the material apple and the reflected light, but not the redness; redness exists in our mind as subjective reality. Space, matter, energy, and time are the elements of objective reality. *Objective reality is external and discrete.*

B. Subjective Reality

Objective reality is nothing more than the configurations of quarks, electrons that form matter, and photons that produce light; everything else that we experience is immaterial. Our minds create subjective reality. Consider the following: the redness does not inhere in the apple; it is associated with the nature (wavelength) of the photons reflected from the skin of the apple. Photons are not red, nor does the redness arise in the eye which brings us to the neurons in the brain. Neurons merge redness with the form of the apple in the brain, but neurons or any combination of neurons are not red. Neurons are discrete, and redness is amorphous (formless). Redness is a subjective experience.

Subjective reality exists in an immaterial memory. The experience of color, of light itself and all the other sensations did not exist until consciousness appeared with the first metazoan. Consequently, the entire universe was silent, odorless, tasteless, and dark for the first 13+ billion years. Neurologists have not arrived at a solution to the *binding problem*, the mental process that allows us to perceive an image of a red apple by combining the form of the apple with its color. The question as to how the discreteness of the form combined with the amorphous experience of the color when we have only the discrete neurons of the brain to work with, cannot been answered. Neurologists can describe the generation of the apple's form as the function of the elements in the eye called retinal ganglion cells. Such cells distinguish difference in intensity between created lines, and the lines create forms. There is no universally acceptable explanation for differentiating a tree's green leaves from a background of a blue sky relying solely on neurons. We can only say that it happens in the brain. *Subjective reality is internal and amorphous.*

C. Rational Reality

An important part of the recognition of the apple sitting on the table is the symbol that we associate with the image stored in the mind, in this case, the word "apple." As I will argue later in more detail, *the mind is a combination of the language faculty residing in the neural map of the brain and a psychical substance (nous) in which the neurons are immersed.* The neurons form a material memory, and the nous forms an immaterial memory.

Symbols are essential for the formation of the mind. Symbols consist primarily of words and numbers, both written and spoken. Symbols also include signals and musical notes from which humans can extract specific meaning. Higher level animals certainly can recognize objects such as apples, but they cannot share that knowledge with other animals; they do not possess the ability to form symbols that are the main components of thought and communication. The brain then is the processor that integrates discrete symbols stored in the *material memory* with the intellect stored in an immaterial memory. Although the brain recognizes and translates symbols, the communication of symbols occurs externally to form rational reality, the collective knowledge, and experiences of humans. We can think of rational reality as a spherical layer of individual (discrete) minds linked through external symbols (language) enveloping the globe, a layer Pierre Teilhard de Chardin called the *noosphere*. Rational reality *is external and discrete.*

D. God

Since the issue that I am addressing is between secular humanism and theism, between the non-believer and the believer, my thesis is based on the existence of a "generic God," a transcendent power acting at the explicative level, described as God, pure existence, and absolute being. As a Roman Catholic, I believe, as most Christians do, that God exists as a Trinity of persons; (i) as a transcendent entity, (ii) as a creator, and (iii) as a personal God—three persons sharing a single divine substance. However, since this is a polemic in which the adversary does not believe in God, I design my argument for use by any believer in God.

1. The transcendent God is the architect of reality and defines the purpose and the rules by which reality functions. The transcendent God is perfect, formless, infinite, eternal, immutable, omnipresent, omnipotent, omniscient, and absolute goodness. He is all that is possible and possesses the power to directly create and sustain all of reality, both the reality we can observe and know and the reality we can never observe or even know. The transcendent God is God the Father.

2. God the Creator is the cause and the prime mover of the universe. God the Creator is instrumental in creating the universe, earth, life, the first human, mind, and soul. God the Creator is finite in the sense of being coterminous with the universe, yet infinite in the sense of being infinitely divisible. God the Creator determines, from within, the timing and sequence of our creation. God the Creator is the Holy Spirit,

3. The personal God interacts transcendentally with man. He is compassionate and is the intercessor of our prayers. He set the example by which we find fulfillment through our involvement in His Mystical Body through the Christian Church. The personal God appeared in the body of Jesus Christ.

The question of how God creates and sustains reality must address how God might exist transcendently, creatively, and personally, i.e., as the Christian triune God: Father, Son, and Holy Spirit. I contemplate the nature of God's transcendence by seeking answers to questions that primarily deal with the nature of physical being (ontology). I contemplate the nature of the creative God by seeking answers to questions that primarily deal with the purpose of creation (teleology). I contemplate the existence of the personal God by seeking answers to questions about our individual lives (theology). My description of a triune God is a religious conviction. Hence, any attempt to counter the materialist, the skeptic, and the cynic with my personal description of a triune God in any discussion with non-believers would be a non-starter. Therefore, I describe a credible generic God that is not easily dismissed with a hand-wave, but which can set the foundation for a later argument for the triune God. I must devise the nature of a generic God that explains reality without conflicting science's descriptive level.

E. Transcendent Reality

Certain words such as beauty, truth, and justice are associated with experiences that do not fit nicely with objective, subjective, or rational reality. I contend that such experiences form a fourth level of reality, a transcendent level. We recognize a transcendental when we engage specific examples. We can agree on the beauty of a sunset or a specific work of art; we can agree on the truth that love is better than hate or that the earth is not flat; we can agree that it is just to remove a murderer from society. Despite that, there is always disagreement as to whether something represents beauty, truth, or justice, but there is a common agreement that beauty, truth, and justice all represent a general transcendental, a primarily desirable experience called goodness.

Despite the inability to explain the nature of transcendentals, we can agree that beauty, truth, and justice exist, that a certain meaning is inherent in such transcendentals. Such ideas exist externally without form but still evoke feelings within. We say that they are transcendent. Beauty, truth, and justice are uniquely human experiences that transcend the objective, subjective, and rational. *Transcendent reality is external and amorphous.*

These modes of reality did not appear together at once, they arrived sequentially. In other words, they evolved. To explain this evolution, I present a path of actualization in figure 3 that represents a plot of material complexity over historic time.

Chapter 11—Path of Actualization (POA)

The Path of Actualization representing the evolution of material complexity in time is depicted by the POA chart in figure 3. Material complexity has increased slowly except for four times when material complexity experienced a rapid increase during the creation of the universe, earth, life, brain, mind, and soul. I refer to these six quantum leaps as creation events: *cosmogenesis, geogenesis, abiogenesis, somagenesis, psychogenesis, and noosgensis.* Before explaining these events, I deal with consciousness, the main factor involved. An idea—found in the writing of Teilhard de Chardin—is that consciousness tracks material complexity; this is implied in the first four creation events. The last two events represent, as indicated by the dotted lines, consciousness separating from material complexity when forming the creation of the mind and soul.

The path of actualization implies that the relationship between consciousness and material complexity began at the big bang. We can trace consciousness back to a time in the sentient stage to argue that the lower the material complexity, the lower the consciousness. We then assume that consciousness, or at least that which is the cause of it, is present wherever there is matter. In the common view, this observation is called *panpsychism*, a convenient abstraction that is presented as an explanation. In my view, spatial dualism explains how consciousness tracks materialism.

Matter, (the physical), is associated with the rational numbers; consciousness (the psychical) is associated with the real numbers. Consciousness appeared only when the first sentient organism appeared, so showing its existence prior to that time is merely a convenient way of portraying the psychical side of reality.

The creation events divide the POA into six stages following the big bang. Each stage began with a creation event. The creation events are associated with new forms of matter. Although there have been attempts, science has not found an accepted explanation for any of the creation events.

As shown in figure 3, following the pre-universe there were 6 stages: (i) cosmological; (ii); geological; (iii) biological; (iv) sentient; (v) sapient; and (vi) spiritual. Each stage begins with: (i) a new form of the matter that evolved during the previous stage or (ii) in the last stage as a quantum leap in consciousness. The new forms of matter that introduce a new stage are the universe; the earth; cellular life; multicellular life; and the mind. Each new form of matter initiates a stasis period of rising material complexity/consciousness. The end point of the POA, the human mind, the structure with the highest complexity of matter in the universe is also associated with the highest level of. I now explain the interaction of consciousness and material complexity

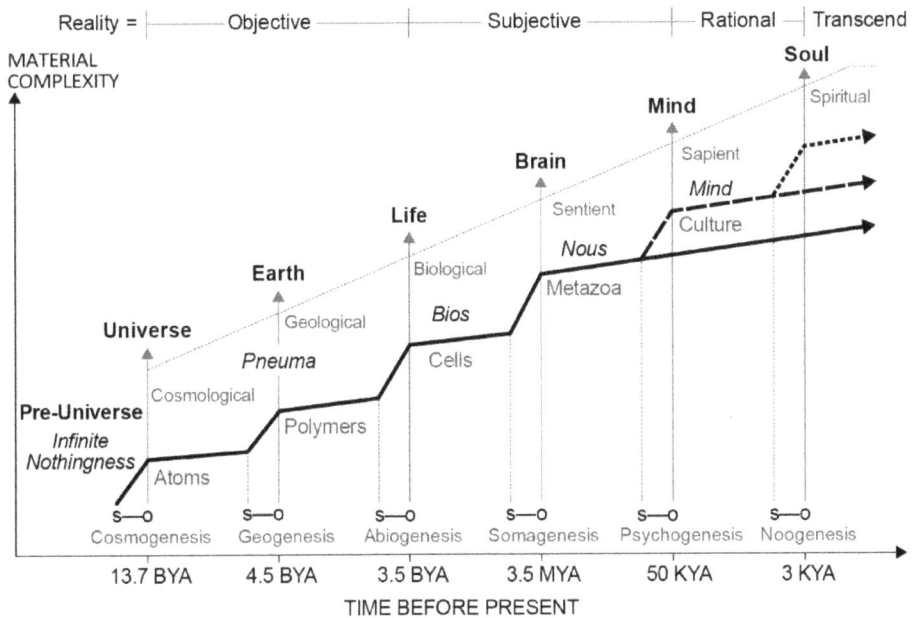

Figure 3. The Path of Actualization

Consciousness vs. Material Complexity

There is an infinite number of real numbers associated with each rational number; when the intensity of the rational numbers generates intensity of the matter, because each rational number is immersed in an infinity of real numbers that generate the complexity of consciousness

> Note 21: Since there is an infinite number of real numbers associated with each rational number, and the intensity of the rational numbers generates intensity of the matter formed by the associated rational numbers that generates the intensity of consciousness. Hence, it is the duality of numbers that induces the duality of space that creates the attachment of consciousness to material complexity.

The forms of matter that represent the first four creation events were *actualized*, i.e., *brought into reality* when noumena are transferred from the realm of potentiality to become phenomena in the realm of actuality. Actualization can be a eureka event or it can be the result of a deliberative process. The actualizations that occurred before the mind appeared were natural, meaning that God did all the actualizations through the foundational path. When natural actualizations occurred, they occurred sequentially, as a kind of probability. In most cases the probability of the actuality of a new form of matter happening depends on what has occurred previously. For example, the actualization by nucleosynthesis of elements beyond carbon depended on the elements created before carbon.

We can address subjective reality more explicitly using hylomorphic duality. In this section I will address the unanswered questions associated with subjective reality and its dichotomies such as mind-matter; immaterial-material; and psychical-physical. Fortunately, there is a substance associated with a plausible explanation that solves the problems of the mind; it, of course, is nous, the psychical substance that is an analog of continuous space.

A. The Stages

The cosmological stage began about 13.6-bya with cosmogenesis, the first creation event. On the chart, the beginning of each creation event is shown by the letter "s" and is ended with the letter "o." The "s—o" is a measure of the creation event duration. For cosmogenesis, the "s" signifies the big bang, and the "o" signifies the formation of the universe when the first stars began to shine. The "s—o" took about 100-milion years to the start of the cosmological stage. Once the stars were formed, there followed a 9-billion-year stasis period during which the path rose slowly. It took three generations of star-formation to create the materials from which the solar system and the planet earth were formed in a 50-million-year creation event called geogenesis. That ended the cosmological stage.

The geological stage began about 4.4 bya with the result of geogenesis and the creation of the planet earth. The geological stage consisted of a billion-year stasis period during which a geological path prepared the surface of the earth, while a biological path developed the polymers from which life emerged. The geological stage ended with abiogenesis, the third creation event, the creation of life.

The biological stage began about 3.6 bya with the appearance of cellular life, the result of *abiogenesis.* It consisted of the biological stasis period of 3-billion years that ended with somagenesis, the fourth creation event, when multicellular life appeared.

The sentient stage began 540 mya with the first examples of multicellular life spread across of 22-23 morphological plans that initiated the *sentient* stasis period during which evolution experimented with slowly increasing material complexity and increasing *cephalization* of metazoa. Subjective reality existed within the crania of these new life forms prior to somagenesis, nothing but the sequential incrementation of configurations of s-points existing in absolute silence and darkness. It ended with psychogenesis when the first Homo sapien appeared.

The sapient stage began with the results of psychogenesis, the first mind, and continued with the sapien stasis period evolving with human culture

B. Interpretation

The path of actualization is based on the observations of science. Science created a compelling description of how reality proceeds along the path. Science's view of objective reality presumes that the action of the fundamental forces of nature acting on matter explained everything. The result is numerous disparate models that describe reality but explain none. I contend that to find a comprehensive view of creation we must take a bottom-up approach and begin at a deeper level of reality (*the explicative level.* We can never be sure of what that might be, but in Part I, I presented a thesis describing the creation of reality at the explicative level by arguing that the foundation of objective reality was discrete space.

My view is based on the premise that God exists, and He creates and sustains at the explicative level that we must explore to find a single mechanism that explains the comprehensiveness and coherence of reality. What I have presented as a mechanism may or may not be true, but it may be plausible enough to open a door to a new direction of the role played by science.

A lifeless universe is a vast and meaningless edifice. Life, especially human life, is what gives meaning to the universe. When dispassionately observed, it appears that our reality as humans, playing out our dramas on a cosmic stage, has arrived at the true purpose of God's plan. We discover what that plan is only after we explain the coherent and comprehensive truth of reality.

If the path of actuality was a smoothly increasing plot of material complexity—a view implied by the theory of evolution—from which we human beings evolved from a common ancestor in small steps over billions of years, consciousness and mind might be easier to explain. However, enigmatic discontinuities have created gaps in science's view at several junctures. Science's position regarding the dilemma posed by the gaps is an imperious proclamation that when we find the answer, it will be based on the fundamental principles of physical science. In the meantime, within the gaps, science cannot offer an explanation or mechanism for the actualization of new forms of matter. Science offers no cause or reason for the big bang. Science offers no plausible cause or reason for how the Earth acquired the set of narrowly formulated conditions that support life. Science offers no plausible cause or reason for the appearance of life and consciousness. Science offers no plausible cause or reason for: (i) the appearance of multi-cellular life in all its vast diversity; (ii) the appearance of the human species and its emergent properties of mind, will, and soul.

Religion offers God as a cause and reason for the gaps. Science scoffs at this "God of the gaps" view, pointing out that as scientific knowledge increases the gaps diminish and so too does God. However, the five major *gaps*—creation events—have not diminished God one iota. The God of the gaps is acceptable to me because it happens to be the position of my religion, namely, that evolution is acceptable if God is the first cause of the creation of the universe, that God created life and the soul, and that God created humanity from a single pair of man and woman. In other words, evolution is acceptable everywhere except in the gaps. What each gap represents is a tremendous increase in the complexity of matter in a brief period of geological time. Imposed on the long smooth curve of gradual evolutionary change, the gaps stand out as major creation events, creation because it is the only way to explain and describe them. It is my intent to describe how the model of dual reality that I described in Part I allows an explanation of the sudden appearance of: the soul.

Chapter 12—Pre-Universe

The most enigmatic philosophical question confronting the human mind is: Why is there something instead of nothing? After the discovery of the big bang, we rephrased that question to be—What came before the big bang? This amazing scientific theory produced a dilemma for the materialist because the big bang theory certainly looks like a creation event that could only have involved God. The big bang theory contends that the universe began as an infinitesimal volume, the singularity, of some unknown substance that expanded explosively to form the universe. I argue that the universe was finite in the beginning, it must still be finite because nothing finite can reach infinity, hence we must add another question to what came before, namely—What lies beyond the universe? In other words what is the nature of the before/beyond, the pre-universe? The believer will simply say that God existed before the big bang, and nothing but God lies beyond.

The materialistic scientist (most of them) will not deny the existence of God directly but instead will find alternative answers even though the answers violate the principles on which science is based. The principle I am referring to is that *"any explanation not verifiable through the senses should be rejected as meaningless."* Well, what happened before and during the big bang is not verifiable through the senses, hence any attempt to explain it is not scientific. The party line among the materialists is to claim there was no beginning. The most direct approach to denying a beginning to the universe is to assume that the universe is infinite, eternal, and consequently hence had no beginning. A scientist that assumes the unobservable, like an infinite universe, is like a theist believing in God. This does not prevent materialists from changing the rules of the game to assist whatever philosophical argument suits their position. Hence, there exists a substantial effort to deny that the universe had a beginning by inventing a variety of highly speculative scenarios that are infinite in extent and time, hence no beginning. This like most other questions divides us between an infinite, eternal universe, or the concept "creatio ex nihilo" creation from nothing.

A. Infinite Eternal Universe

The most direct materialistic approach to the question of what existed prior to the universe is to assume the universe is infinite, that nothing preceded the big bang, that it had no beginning or, given the reality of the big bang, that the universe was only part of a multiverse with an infinite number of big bangs. Scientists assume the universe is infinite because the space that determines the dimensions of the observable universe appears to be flat. Anything travelling in flat space must travel in a straight line and the only end for straight travel is infinity. This, of course, ignores the possibility of the two modalities of space that form a border.

One-way cosmologists address the beginning problem is to suppose that the universe emerged from an energy-fluctuation that formed a bubble in rapidly inflating space that began in the same way in a much larger universe. When this scenario is repeated ad infinitum, there is no beginning. It assumes that inflation was eternal, and the number of bubble universes is infinite. And here we run into a vicious infinite regress. Another proposal was that the expansion of the universe would eventually reach zero at which point it would contract back into a singularity in a 'big crunch' only to undergo another big bang. This cycle has been going on and will continue to do so forever.

String theory and its multiple dimensions joins the game with the idea that the universe's 3-dimensions emerge from higher 11-dimensional one of an infinite number of worlds called branes. Then there is the Many-World Interpretation of the Schrödinger equation that suggests that each time we are confronted with a choice between two options, we are choosing between two independent worlds, both of which exist, hence parallel worlds. Multiverse theories are not science fiction per se because there is plenty of wiggle room in science to utilize the laws to derive models describing some fantastic scenarios just mentioned. Such hypotheses are derived (by mathematical formulation) from modern scientific theories such as the general theory of relativity, quantum mechanics, and string theory. Given the complexity of scientific equations many scenarios are possible. Given that the observed evidence for the big bang theory makes any other scenario completely speculative and strengthens the plausibility that the big bang is a creation ex nihilo event. The materialist must revert to plausibility as the criterion for his/her hypotheses, and plausibility puts my thesis on a level playing field. As does describing the nature of singularity!

B. Creatio ex Nihilo

The *big bang theory* contends that the physical elements (space, matter, energy, and time) emerged from a singularity, an unimaginably small particle of an indescribable substance to construct the entire universe. This scenario creates two questions: (i) *what was the nature of the singularity*, and (ii) *what was the nature of the pre-universe*? The big bang theory proposes that the universe had a beginning in time and thus supports for the Catholic doctrine of "creation from nothing." The big bang theory strongly suggests a finite universe, which, in turn, suggests a boundary separating the universe from what existed before and beyond; we believers would argue that the before/beyond is the Mind of God. The only plausible description for what came before the big bang and still lies beyond the universe is pure existence and/or absolute being, that I refer to as *infinite nothingness*. Here I am playing with words in that I do not equate nothingness and nothing. I use the word nothingness to mean "having the appearance of being nothing." As argued earlier, the negative interpretation of "nothing exists" is an impossibility, and therefore nothing (nihilo) must have an existence as infinite nothingness the only imaginable possibility.

Since the big bang theory contends that not only space but also time, matter, and energy emerged from the singularity, the impetus that initiated the expansion could not have been energy-fluctuations that science describes as the impetus for the big bang. The only plausible interpretation is that energy did not precede the big bang and could not have been the impetus. Since my basic premise is that God exists then surely the big bang must have been a God induced creation event.

There are 100-billion galaxies organized in clusters, super clusters, and galactic filaments. An average galaxy contains 100-billion stars. Our sun, an average size star, is an immense ball of hydrogen converting to helium at millions of tons per second. The incredible numbers of enormous objects represent a fantastic amount of matter that emerged from an infinitesimal object called the singularity. How was it possible that the total mass-energy of the universe squeezed into an infinitesimal object? Science has no definitive answer, but that does not prevent cosmologists from offering speculative opinions.

Science has demonstrated that matter is interchangeable with energy, so it is not implausible to accept the spontaneous creation of matter from a concentration of energy in a background of space, space organized as a field called a false vacuum. The current hypothesis is that something called the Higg's field creates mass, hence matter. Science has no problem attributing magical properties to space, especially regarding objective reality. Did space emerge at the big bang as is the common view? How was gravity created if there was no space and hence no gravitational field? Could gravity and the other fundamental forces emerge from non-existence? I do not think so. Consequently, the statistical accident explanation does not hold water. Something else was involved. Something else gave us gravity, light, and magnetism and endowed *space* with its magical properties. In other words, something else created and organized what we observe, using rules that we recognize as the laws of physics.

Although I use the word *nothingness* to describe the pre-universe, whatever it was it certainly was not a void. Whatever came before and exists beyond the universe, the before/beyond, is pure existence that possesses definable properties. For example, it certainly possessed the cause of the big bang. Where else could the laws of physics reside? Math? Logic? The impetus for change? I contend, and as I will argue in later chapters, qualia, feelings, meanings, concepts, and emotions have their source in the substance of the before/beyond, the thing I call *infinite nothingness*, a psychical substance. In fact, the one indisputable property of the before/beyond is that it contained all that is possible—a realm of possibility—from which reality emerged in a sequence of actualizations. Whatever is possible exists somewhere, somehow. Only the impossible does not exist. Hence, we can say that what existed before the big bang, the pre-universe is pure existence, absolute being, it is the infinite nothingness, the before/beyond, the realm of possibility, the Mind of God.

C. Cosmogenesis—Common View

The big bang theory is science's explanation for the creation of the universe. It was developed by extrapolating back from our current expanding universe to the singularity using empirical knowledge of the relationship between matter and energy. If we heat (add energy) a solid eventually becomes a liquid. As we add more energy, the liquid becomes a gas. At a much higher temperature, the atoms dissociate to become ions and electrons—the gas becomes plasma. Beyond the plasma point, the application of heat cannot supply enough energy to affect changes in matter; hence, physicists use particle accelerators to impart enough energy to generate a transformation. Information obtained from crashing highly energized particles into one another allows physicists to describe the interaction of matter and energy back to the first 10^{-43} seconds of the big bang, the so-called Planck era, during which, the universe was a perfectly symmetrical homogeneous substance in a simple state and was acted upon by a single force. At 10^{-36} seconds, the strong force separated from a single composite force to start the electro-weak epoch during which the weak force remained merged with the electric force. Also, at 10^{-36} seconds, the universe began the inflationary period of rapid expansion during which the universe doubled in size a trillion times. Inflation lasted about 10^{-32} seconds.

After the inflationary period, the universe expanded at a much slower rate for 10^{-12} sec when the weak force separated from the electric force and the homogeneous substance granulated to form a plasma of leptons, quarks, gluons, and their antiparticles. At 10^{-6} seconds, quarks formed protons (the hydrogen nucleus). At 1-second, leptons annihilated their anti-partners; the resulting release of energy created vast numbers of photons. At 3 to 20 minutes, helium nuclei formed through big bang nucleosynthesis. This epoch lasted for 380,000 years after which the photons decoupled from the atoms and universe became transparent. The big bang ended and for the next 100 million years the universe remained *dark* until the first stars began to shine. However, there was no mental faculty to experience the stars radiant show, so the stars were playing in an empty stadium and the universe technically remained in darkness until the first sentient creature with eyes experienced the *subjective experience* called light.

The "*surface of last scattering*" (SLS) represents the decoupling of the photons. Its significance is that we cannot see back into the first 380,000 years, so when we speak of the size of the universe, because inflation took place within the SLS epoch, we do not know the size of the actual universe, but Alan Guth postulated the inflation theory to resolve the following problems associated with the original theory: (i) the absence of relic particles; (ii) the horizon problem; and (iii) the flatness problem.

Relic particles are particles predicted theoretically to have existed at the very beginning of the big bang. These include, for example, unobservable magnetic monopoles and gravitons. In attempting to solve the monopole problem, Alan Guth proposed that in the first moment of the big bang the universe expanded astronomically, and the monopoles and other relic particles dissipated over the vast universe resulting at such a low density that they are undetectable.

The horizon problem arises because observations, in all directions, have shown that the cosmic background radiation appears homogeneous; this is a dilemma since measurements show that the distance across the universe from one end to another is greater than the distance light can travel. The age of the universe is 13.7-billion years and the distance to the edge of the observable universe as defined by the *surface of last scattering* (SLS). The best estimate of the distance to the SLS (the radius of the observable universe) is about 46 billion light years. Because the outer edges of the universe are beyond the distance light could have travelled in the lifetime of the universe, radiation cannot distribute uniformly. Before inflation, all parts of the universe were in contact. Also, radiation was uniformly distributed, then inflation locked in uniformity resulting in a homogeneous and isotropic universe.

The flatness problem relates to the curvature of space. According to the general theory of relativity, the curvature of space depends on the proximate mass-energy density. Application of the general theory to the universe requires a curvature within 1% of flatness (zero curvature). The equations also predict that space is flat at the beginning of the big bang. Initial observations of flatness did not meet the theoretical calculations. In a perfect example of mathematical formulation and abstraction, the inflation theory was introduced to resolve the problems.

D. Cosmogenesis—My View

The only plausible scenario implied by the big bang theory is a single finite universe created from nothing. I agree with the details, given in the previous section, of science's description of the creation of the universe. I differ from the scientific view of an infinite universe. With an infinite universe, science has no need deal with a creation event that strongly suggests the existence of God.

In Chapter 1, I pointed out that: (i) the universe is finite; (ii) that it emerged from a singularity; (iii) it has a border composed of discrete and continuous space; (iv) what came before and exists beyond the universe is the Mind of God that can only be imagined as pure existence, absolute being, and as I have argued previously, infinite nothingness.

The equivocation that "the universe is infinite" can only mean that "the multiverse is infinite" and for that there is no evidence. The only thing on which we can agree is that only one universe appeared suddenly from the before/beyond and expanded to what we observe presently; everything else is speculation. The observation of the universe's expansion does not support the multiverse hypotheses; they are the product of speculative mathematical formulation.

The question that science avoids is how did the vast amount of matter found in the universe fit inside the infinitesimal object called the singularity? Science has no way of describing its nature. On the other hand, if as I contend, the space that emerged from the singularity to form the space that defines the universe is discrete space, then we can explain the nature of the singularity as follows. Discrete space consists of s-points that are dimensionless, and God was able to squeeze as many s-points into a volume as small as we would care to make it. Since matter consists of configurations of dimensionless s-points, there is no problem squeezing all the incipient matter that formed the universe into an infinitesimal singularity.

E. The Singularity

Science has no explanation or makes no attempt to explain a singularity. However, in my view, because the universe consists of discrete space consisting of s-points separated by s-gaps, the size of the universe is determined by the number of s-points and the size of the s-gap. Therefore, since a point is dimensionless, simply by decreasing the s-gap to an infinitesimal distance, like 10^{-1000} meters or some such separation, an unlimited number of s-points will fit into an infinitesimal volume. At the present, the s-gap equals the Planck length, $(l_p) \sim 10^{-35}$ meters; but when we run the clock backward in time, we find the s-gap was decreasingly smaller. Since the width of the s-gap determines the size of the universe and the speed of light, it must have approached zero in the singularity. Also, we do not need an infinite number to construct the cosmic s-frame that is 92-billion light years in diameter. Assuming an average spacing between s-point of $\sim 10^{-35}$ meters, a simple calculation yields about 10^{+183} s-points are required to construct the universe. This finite number is the root of the conservation laws that were discussed in Chapter 8.

1. Initial Impetus

The second question looking for a final answer is: what caused the universe to suddenly expand? The biggest roadblock for any answer to this question is the law of conservation of matter and energy that dictates that all matter and energy must be present in the singularity. The common view contends that the energy of expansion was cancelled by the negative energy of gravity resulting in zero starting energy. Okay, but what initiates the big bang? A quantum fluctuation? Perhaps! I prefer something much simpler.

My basic premise is that God exists and the mechanism He used to create the universe was the holonomic mechanism, a cellular automaton for which the initial configuration of s-points was in the singularity, which when acted upon in accordance with a cosmic algorithm, and given the impetus of incrementation, resulted in the vast universe we now observe. In Chapter 5, I described an s-particle as a fixed number of coincidental s-points piled up at a single mass point enclosed by pneuma. The singularity is a cosmic s-particle, the mass of which is determined by the number of s-points in the cosmic mass point. The s-gaps in the singularity were infinitely closer than $\sim 10^{-35}$ meters, the size of the present gap. Since the Mind of God supplies the initial configuration and the impetus that increments the cosmic s-frame, universal space, energy, and time began with the expansion at t_0 — matter came later. Hence, space, matter, energy, and time emerged from the singularity with the big bang as science describes. The expansion that has come to be known as the big bang, which at the descriptive level is an indescribable explosive like expansion of near infinite proportions, is at the explicative level, a simple expansion of the s-gap. No energy or unknown force is required only the application of the holonomic mechanism for which the cosmic algorithm determines the width of the s-gap and God provides the impetus of incrementation. I assume that the non-believer would find laughable the idea that the infinite nothingness, the Mind of God, has the capability to produce what is to become an object of enormous mass and unimaginable energy. But at the explicative level, there is no energy, just an enormous number of massless s-points in a singularity; the energy comes later. God needs only to increase the size if the s-gap to create the incremental change in the cosmic s-frame from which energy and time emerge as the universe expands. Increasing the s-gap requires no energy because s-points have no mass.

Note 22: The fixed number of s-points in the singularity that eventually form matter and energy restrict the amount of mass-energy, hence the conservation of mass-energy.

Chapter 13—Cosmological Stage

At the beginning of the 20th century the cosmological paradigm had been unchanged since Copernicus, Galileo, Kepler, and Newton explained the motion of the stars and planets and replaced the Ptolemaic system by a static universe that existed without a beginning and consisted of a single galaxy in which the planets circled the sun. Long before the invention of the telescope, astronomers observed interstellar patches of gas-like luminosities called nebula. They thought that nebulae were closely bunched with a multitude of stars. As telescopes improved, it began to appear that nebulae were gaseous, and a debate ensued during the first part of the twentieth century that was only resolved by a series of improved observations that also led to the observed expansion of the universe and the big bang theory.'

The cosmological stage began with the universe that was formed during cosmogenesis as described by the big bang theory and extends through the cosmological stasis period. This stage began with a sudden increase in material complexity from a simple singularity followed by 9-billion years of slower complexity growth. This stage featured a geologically brief and violent interaction of the physical elements—space, matter, energy, and time—followed by a long cooling down stasis period during which the universe evolved in preparation for the next creation event. It was a period of physical activity, of energy driving matter through space and time.

A. Cosmological Stasis Period—Common View

When the big bang ended after 380,000 years, the universe was a vast expanse of hydrogen with lesser amounts of helium and some traces of lithium and beryllium spread out as a cosmic molecular cloud. There were a vast number of fluctuations of minutely greater density distributed throughout the cosmic molecular cloud. As the universe cooled, the fluctuations that had grown as the result of inflation to galaxy sized molecular clouds continued to aggregate matter until they collapsed and formed the workshops in which the physical elements produced the structure and material to create the earth.

There is no established theory of galactic formation, and whether the galactic formation was either bottom up or top down. Either way, the predominant hypothesis is that: as the molecular cloud collapsed, the temperature of the high-density regions increased to form rotating spheres of superhot gases creating the matter that formed protostars. The protostars continued to aggregate matter from the surrounding molecular cloud until the internal pressure raised the temperature high enough for hydrogen to fuse and form helium and a star was born.

In a typical star, when the heat generated by the fusion provides an outward pressure that counterbalances the pull of gravity, the star reaches thermal equilibrium and begins a long period of hydrostatic stability. The star remains a stable main sequence star until it consumes all the hydrogen in the core and the star evolves to a different type. The type of evolution depends on the mass of the embryonic protostar from which it emerged.

For a star, the size of the sun, a yellow star, the depletion of the hydrogen core results in a decrease in the outward pressure and the star collapses due to gravity. The heat generated by the collapse causes the hydrogen fusion to migrate to the outer layers and the star becomes stable again. Eventually the outer layer of hydrogen is depleted, and the star collapses around the core generating enough heat to cause ignition of helium. The helium burning initiates *stellar nucleosynthesis*, the fusion of lighter elements to form heavier elements. Nucleosynthesis ends with the formation of carbon and oxygen in stars with smaller masses than the sun. If the heat generated by the helium burning is greater than that needed to counter-balance gravity, the small star expands and becomes a red giant then collapses to become a planetary nebula, whose core becomes a white dwarf and eventually, as it cools, a black dwarf. For stars with heavier mass than the sun, the embryonic *protostars* evolve along a similar sequence of collapse, nucleosynthesis, expansion, and core and outer layer fusion. However, the large mass star forms heavier elements beyond carbon and oxygen and becomes a red super giant that eventually explodes as a super nova, then collapses to form either a black hole or a neutron star.

The universe resulted in a variety of galactic types that appeared during the earlier generations using different processes that are not well understood. Galactic formation, movement, and collisions still occurred throughout this stasis period resulting in the formation of a large-scale structure as galaxies clustered in sheets and filaments outlining vast volumes of space devoid of luminous matter. Galaxies are separated by enormous distances measured in terms of a million light years. Despite the tremendous amount of matter contained in the galaxies, the universe is mostly empty space in which the luminous mass in the stars represents only 5% of the total mass of the universe. It took about 9 billion years for at least two previous generations of super nova stars to seed the galactic molecular cloud with enough heavy elements before geogenesis could occur. The first-generation stars were short-lived (a million years) and massive (300–1000 times the solar mass); they were efficient producers (through nucleosynthesis) of heavy elements. The second-generation stars also contributed heavy elements to the galactic cloud. Our sun, a third-generation star, was formed from planetary nebulae rich in heavy metals. How amazing is our God to have taken three generations to patiently produce the necessary materials to create the earth, then to leave behind for our edification those expended furnaces as jewels in our night-time sky? The interstellar galactic cloud, initially was primarily hydrogen and helium, became increasingly granular as stars formed, died, and contributed light and heavy elements, and molecules of water, and carbon compounds to the interstellar galactic cloud. As a result, after 9-billion years the galactic cloud was finally ready to activate the next creation event. Considering the vast number of stars in the vast number of galaxies, the chances would be great enough, for a star of just the right size, to have circling about it at just the right distance, a planet formed from just the right materials, organized in just the right structure, to allow sustainable life to form. Such fortuitous conditions were certainly possible and even probable for geogenesis to occur statistically, but when we combine that probability with all the other improbabilities, the view of creation leans heavily in favor of the work of God. But that is the argument forwarded by the intelligent design guys and that argument has been hand-waved out of favor by the secular humanist gang. That is one reason for my polemic, to give the Intelligent design guys a description as to how God intelligently designed total reality.

B. Cosmological Stasis Period—My View

At the descriptive level of the big bang, fluctuations in the molecular cloud collapsed and fragmented to form the seeds of the individual stars. It is hard to imagine how a slightly denser gas of hydrogen would collapse into a proto star especially since the hydrogen atoms are receding from one another because of the expansion of space. It is no harder to believe that stars form from the aggregation of molecular clouds, than to believe that the entire universe fits into a volume smaller than the period at the end of this sentence. But the transition of slightly dense hydrogen collapsing against the expansion of the universe is science's "most plausible" hypothesis. However, plausibility is not always the end of explanations. Since Infinite Nothingness extends to infinity, the volume of the singularity—or as I will call it *"the cosmic s-particle"*—is finite and its mass is equal to that of the universe. Implied in this view is the idea that the more s-points coincidental at a mass point (the singularity), the more massive the spatial particle is. Enough points can congregate in an infinitesimally small volume such that there is enough incipient mass in the singularity to construct the universe. Since an s-point takes up no space, an infinite number gathered at infinitesimally small volume suggests the nihilo in creatio ex nihilo.

God has arranged the s-points within the singularity as a telic configuration. Theno incrementation according to the cosmic algorithm will result in the universe as it appeared at the surface of last scattering—the distance travelled by the photons emitted from the cosmic background radiation today. The number of s-points concentrated at a mass point range from the total mass of the universe, the cosmic s-particle, to the smallest s-particle, something that is barely more than pure space. Intermediate between a s-particle of minimum mass and the cosmic particle are s-particles with masses of all intermediate sizes. The stuff that spewed from the cosmic particle in the first 10^{-43} seconds of the big bang was not homogeneous. Imbedded in and carried along with the immense number of individual s-points, which expanded to form the network of discrete space—to fill the volume of the physical universe—were immense numbers of s-particles of many sizes. In addition to individual s-particles destined to become the missing dark matter, the expanding volume of discrete space also included galactic s-particles with enough s-points to form galaxies. Each of the galactic s-particles may have undergone its own mini-bang to provide matter to construct the galaxies from within. In this way, the universe has become an inhomogeneous distribution of matter and space. Moreover, it happened through the holonomic mechanism, which means that the singularity contained an initial configuration that included innate lumpiness that became the galactic s-particles.

Sounds like science fiction but remember the initial premise that "God exists" and arranging 10^{183} s-points so that they eventually configure themselves in the form of the planet earth is an example of omniscience plus omnipotence and certainly within the capability of an infinite God. My main point, however, is not to claim that this is how the big bang occurred but merely to show that science is not the only plausible path to the truth.

C. Geogenesis—Common View

There are hypotheses describing the formation of the solar system, but because it provides the best explanation for the distribution of angular momentum, the nebular theory is the one most generally accepted. The sun with 99.9% of the mass has less than half of one percent of the angular momentum of the solar system. Jupiter, despite accounting for a small percentage of the total mass, accounts for 99% of the total angular momentum because of its great orbital velocity. The nebular theory proposes that the solar system formed as the result of a gravitational collapse of part of a galactic molecular cloud that formed a flattened rotating disk of molecular gas surrounding a highly dense central region that became a protostar, our sun. The material left over after the protostar formed continued to rotate about the sun as a flattened nebular disk of gas, particles of ice, and molecular dust including carbon-based molecules. The fragments of matter began to form larger and larger solid objects called planetesimals. Once the planetesimals form, they accreted gravitationally into bigger chunks of material called protoplanets that underwent internal melting to produce a differentiated interior. When the object that became the earth accreted

74

enough matter, it began to heat—either through collisions with cosmic debris or from radioactive emissions—and became molten. The constituent elements segregated radially; the heavier elements migrated to the center and the lighter ones to the surface, and slowly cooled to form the core, mantle, and crust.

D. Geogenesis—My View

There is no need for me to argue the miniscule probability of the earth forming *naturally*, the Intelligent Design community has done that with great fervor. The probability argument merely negates the materialist's argument that objective reality occurred *stochastically*. The materialists' counter argument is that the only positive argument Intelligent Design proponents offer has been that "God did it." My approach to countering the materialists' view is not to argue the improbability of life occurring stochastically, but to supplement the God-did-it retort with a plausible argument as to how God did it. In other words, I am offering a positive argument.

I have already introduced the holonomic mechanism as the tool with which God implemented cosmogenesis and sustained the universe through the cosmological stasis period; I now invoke the holonomic mechanism again to explain geogenesis. I realize that this sounds just like another way of saying, God did it, but this must be the case since I believe that at the explicative level God needs only one mechanism. Hence, I use the same holonomic mechanism to show how "God did it." I argue that there is only a single mechanism for explaining multiple phenomena if we are to explain a comprehensive reality. Based on the assumption that God exists, a single mechanism approach is innately comprehensible. The holonomic mechanism that I have presented may not be the precise comprehensive mechanism, but it is general enough to be plausible. Its diversity is in its algorithms. The algorithm for the creation of galaxies and the stars needed modification to create the improbable combination of events required to form the earth. My claim is that God could at any point introduce a telic configuration in the path of incrementation that resulted in the sequence of events, at the precise location for geogenesis to occur. The holonomic mechanism consists of three elements: initial configuration, an algorithm, and an impetus. Based on the premise that God exists, I argue that the impetus for incrementation makes sense when attributed directly to God's omnipotence or at least makes as much sense as attributing the attraction of matter to the law of gravity and not attributing its inexplicable source, namely why matter attracts other matter. Science does not explain gravity; science merely describes it.

Chapter 14—Geological Stage

The geological stage began with the result of geogenesis, the creation of the planet earth, and extended through the geological stasis period. This stage consists of a sudden increase in material complexity that emerged from a cloud of hydrogen, water, and other molecules followed by a billion years of slower complexity growth. This stage features a geologically brief and violent interaction of the physical elements—space, matter, energy, and time—during the formation of the earth, followed by a long cooling down stasis period when chemistry took over to form the complex hydrocarbons from which life emerged. It was a period during which vast experiments into the actualization of complex molecules occurred. It took about 9-billion years for at least two previous generations of stars to seed the galactic molecular cloud with enough chemical elements such as heavy metals, water, and carbon molecules before geogenesis occurred with the creation of the planet earth. Considering the vast number of stars in the vast number of galaxies, the chances would have been great enough, for a star of just the right size, to have circled it at just the right distance, a planet formed from just the right materials, organized in just the right structure, to allow sustainable life to form. Such fortuitous conditions were certainly possible and even probable for geogenesis to occur statistically, but when we multiply that probability by all the other probabilities, the view of creation of the earth leans heavily in favor of the existence of God.

The common view assumes that soon after the sun formed all the planets including the Earth also formed together in a short geological time. Geogenesis, the creation of the earth, initiated the geologic stage about 4.6 bya. According to the *giant-impact hypothesis,* sometime before 4.5 bya the debris left over from a collision of the earth with an astronomical body the size of Mars formed the moon. The moon is essential because it causes the ocean tides that scientists believe were necessary for the creation of life. The moon also provides a stabilizing force on the earth's rotation. Even more important, the moon may contribute to the motion of the core that results in the earth's magnetic field.

A. Geological Stasis Period—Common View

The sun formed about 4.6 bya and the earth formed about 50 million years later after which the geological stasis period that followed geogenesis took a billion years to provide the conditions and materials from which life emerged. It is almost impossible to determine when abiogenesis occurred, but evidence suggests that it was sometime shortly before cyanobacteria appeared. Cyanobacteria use a process called photosynthesis that uses light to convert carbon dioxide and water into carbon hydrate and oxygen. It is the effect that oxygen had on the fossil record that verified the presence of cellular life at a geologically brief time after the earth cooled enough to form rocks. Once the basic layered structure of the earth was in place, the path of actualization proceeded along a dual path, geological and chemical. The former followed the evolution of the physical conditions of the earth's surface and atmosphere while the latter followed the evolution of the molecular structures that emerged in abiogenesis, the creation of the first cell. The paths, of course, evolved as parallel paths that often intersected the geological affecting the chemical and visa-versa. The answer to the question as to what kind of conditions and materials life requires provides a framework for describing what may have happened during the geological stasis period.

The free O_2 produced by the cyanobacteria reacted with dissolved iron until the cyanobacteria began colonizing as large multicellular mats and the increased production of O_2 hastened the iron reaction until all the available iron oxidized. The result was a significant increase in the percentage of O_2 in the atmosphere, known as the *great oxygenation event* (GOE), it led to: (i) a mass extinction of obligate anaerobes; (ii) the multimillion-year *Huronian glaciation*; and (iii) an enhanced diversification of life. The table in figure 4 outlines the geological and chemical paths throughout the Precambrian era leading to the creation of multicellular life.

1. The Chemical Path

While the geological path evolved, chemicals were evolving along a separate path of slowly increasing material complexity. Unlike the hypotheses proposed to describe geological path, which is based on evidence extracted from geological studies, the hypotheses proposed to describe the evolution of chemical evolution are speculative. There has been general agreement that the sequence of events must have occurred in a cosmologically short time of 50-million years. Life appeared around 3.6 bya although there is chemical evidence that it may have appeared as early as 4.1 bya. I use the 4.1 date to represent the seminal event in the POA chart.

The evidence for the early appearance of life is based on the analysis of the graphite form of carbon found within indestructible crystals of zircon from the Jack Hills in Western Australia. The low ratio of carbon 13 to carbon 12 found in the zircon is characteristic of organic processes. It is merely an indication, not proof. How would science explain the likelihood of life showing up about 3.8 bya before the water had formed into pools as supposed? When the first life appeared on the earth, it was a cell without a nucleus (*prokaryotic bacteria*). Based on genetic research the current theory contends that all living organisms descend from a simple bacterial cell. Although bacteria are among the most successful life forms, adapting to larger variations in the environment than the more fragile organisms of the *eukaryotic* world, one must wonder, if bacteria are so successful and "survival of the fittest" is the impetus for evolution, why did life ever progress beyond the bacterial level? In observing the history of life, one must assume that progress had something to do with increasing complexity. There is no sure way to measure complexity, but based on a subjective assessment, the complexity of matter made an enormous jump when the first animated cell and life appeared on the earth. Life ushered in a new stage in the complexity of matter with a structure *described* by modern science, but which defies *explanation.*

B. Geological Stasis Period—My View

The descriptive level of the geological stasis period that provided the foundation for abiogenesis is one filled with scientific speculations. Because there is no way to observe the mechanisms and processes required for life to emerge, the proposed mechanisms can never be more than speculation. Regarding creation, science can only present hypothesis based on plausibility and admit that one mechanism was not involved in the creation of life is the theory of evolution. During the stasis period that paved the way for abiogenesis, DNA, RNA, protein synthesis, locomotion, metabolism, replication, intercellular membranes, and other innovations and mechanisms emerged and combined in a specifically arranged object, the first living cell. And if we are to ascribe to a Darwinian explanation, the new actualizations must have happened in a gradual sequence. What that sequence would look like would have to be more detailed than any currently proposed by science. Since science cannot observe prebiotic phenomena, and speculation is the only option to describe the situation, then plausibility is the only criteria for acceptance as a theory. Plausibility puts the intellectual battle between materialists and believers on an equal footing. What involvement does God have in preparing for the emergence of life? During the cosmological stasis period, I contend that God was involved only at the beginning because the holonomic mechanism sufficed to generate the sequence of increasing complexity that brought matter to the level of protocells. The deterministic laws of physics work well at the macroscopic level of reality but not at the microscopic level. At the microscopic level, the contingency built into the evolution of matter requires a probabilistic approach. Consequently, algorithms are more suitable than mathematical equations for describing probabilistic situations.

In addition, abiogenesis could not have happened without binary fission, the process by which single cells divide to produce two daughter cells. As a cell grows, it reaches a size that triggers a duplication of the DNA. Each duplicate DNA moves to a different part of the cell membrane which then separates into two complete cells. Binary fission is a cellular mechanism that depends on the cell structure and the propagation of the cell structure depends on binary fission. The first animated cell needed a mechanism such as binary fission for duplication or else evolution would have ended before starting. Here we have another "what came before, the chicken or the egg conundrum?" I contend that God's intercession introduced binary fission by inserting a telic configuration into the holonomic mechanism. Yes, this is God in the Gaps, but with a mechanism that explains the gap. I contend that it is easier to write an algorithm than an equation to define when and how fission first occurred.

1. Abiogenesis—Common View

During the 1920's, the Russian biochemist, Alexander Oparin (1894–1980), and the British scientist, J.B.S. Haldane (1892–1964), speculated about prebiotic evolution. Haldane suggested that life would require special proteins. Eventually the complex molecules combined to form the first protocells and eventually the first true living cells. Oparin proposed a theory outlining the creation of life from matter without the presence of a non-material substance.

The argument that life requires a non-material substance is called vitalism. Vitalism is a metaphysical theory that argues that life is due to a psychical agency that is distinct from material structure; consequently, matter is necessary but is not sufficient requirement for life to occur. The two most famous proponents of vitalism were the German biologist, Hans Driesch (1867–1941), and the French philosopher, Henri Bergson (1859–1941). Vitalism presumes that an immaterial substance was a prerequisite for the creation of life. Most biological scientists debunked vitalism based on the philosophical principle of scientism.

One can never know how (and exactly when) life emerged from inorganic matters in the distant past; we can only surmise the nature of abiogenesis. Certainly, we assume that the laws of physics and chemistry were involved. However, there is nothing in the laws of physics and chemistry that predicts or explains the animation of the first cell. Material structure alone does not explain the initiation of the dynamic interaction of those elements that causes them to sustain and replicate. What science offers is an extensive description of a living cell, its *sustenance*, and its *propagation*. For example, Oparin proposed that the creation of life from inorganic matter occurs in four steps during the geological stage of the earth's history; the steps are:

Era	Time	Geological	Biological
Hadean Eon			
	4.60	Formation of the Earth	
	4.50	Formation of the Moon	
Archean Eon			
Eoarchean	4.00	Crust Formed	
	3.80	Heavy Bombardment Ends	
	3.70	Oldest Rock	
	3.60	Oceans Formed	
Paleoarchean	3.50	Supercontinent Vaalbara	Cyanobacteria, O_2, Stromatolites
Mesoarchaean	3.20	Supercontinent Ur	
	2.90	Pongola Glaciation	
	2.80	Vaalbara Breask / Mag Field	
Neoarchean	2.70	Supercontinent Kenorland	Photosynthetic Cells
Proterozoic Eon			
Paleoproterozoic	2.50	Continents Separate / Ocean O_2	
	2.40	Oxygen Catastrophe	
	2.30	Huronian Glaciation	
	2.20		Cellular Respiration
	1.80	Supercontinent Columbia	
Mesoproterozoic	1.60	Columbia Breaks Up	1st Eucaryotic Cell
	1.40		Last Common Ancestor
	1.20	Supercontinent Rodina	Sexual Reproduction
	1.00	Start of Ice Age	Multicellular Organisms
Neoproterozoic	0.85	Snowball Earth	
	0.64		Ediacaran Fauna

Figure 4. Pre-Cambrian Eon

1. Synthesis of monomers of inanimate material,
2. Polymerization, formation of complex molecules from monomers.
3. Formation of protocells,
4. The creation of self-replicating cells from protocells.

This sequence describes an advance of complexity, and it represents an orderly increase of growth which for the first three steps is amenable to a Darwinian explanation. The fourth and last step involves an innovation not explained by a Darwinian evolution. The first cell must have gone from a specific structure of inanimate matter to a living cell. There is no in-between, partially alive structure; life had to suddenly emerge with the final infusion of animation added to a specific structure.

In the 1950's, the American chemist, Stanley Miller (1930–2007) under the guidance of the American physical chemist, Harold Urey (1893–1981), did an experiment in which they applied electric sparks to a mixture of water, ammonia, hydrogen, and methane and produced traces of amino acids, the building blocks of life. The Miller experiment was broadcast throughout the media as proof that life can materialize spontaneously from inorganic molecules. The Miller experiment became a main weapon used to debunk any argument about the possibility of vitalism or any other theory of dual nature of reality. Materialists quickly proclaimed that the Miller experiment was proof that life was an emergent property of matter. Even though the only thing Miller's experiment proved was the first step in Oparin's hypothesis of abiogenesis, it was sufficient evidence to kill vitalism.

However, many biologists realize that the Miller experiment was not the final statement in the materialist argument because it did not provide a mechanism for the last three steps in the *Oparin hypothesis*. Furthermore, the conditions of Miller's experiment produced other compounds in higher concentration than amino acids; that would have prohibited the formation of polymers. And even if they did form, the water used in the experiment allowed hydrolysis, the chemical decomposition of the resulting polymers into their constituent monomers faster than the condensation of monomers into polymers.

There has been more than one hypothesis offered to close the explanatory gaps between the formation of amino acids and the creation of a living cell. These proposals address specific steps in the abiogenesis process without tying them all together with a single coherent mechanism. Although there is no direct evidence that the laws of physics and chemistry describe the sequence of Oparin's first three steps together, there are plausible scientific descriptions of processes that separately implement each of the first three steps. Although the Urey-Miller experiment, by creating amino acid from laboratory simulated early earth conditions, verified the synthesis of monomers (the first step), none of the other proposed processes for the evolution of polymers and protocells have gained general peer-approved acceptance. And even if found, the most difficult barrier to hurtle, step 4, the creation of self-replicating cells, would still exist. Much work has been done searching for a solution to the mystery of life. For example:

1. The deep-sea vent theory, radioactive beach theory, and Gold's deep, hot biosphere theory address conditions.
2. The clay theory, the iron-sulfur world theory, polyphosphates, and the lipid address the polymerization challenge.
3. Fox's microspheres and bubbles address the protocell challenge.
4. The scientific community cannot agree on what came first, DNA or RNA,
5. and to jump right to the solution there are those who argue that life originated in outer space and came to earth on the back of a meteorite.

The origin of life on Earth is a set of paradoxes. For life to emerge, there must have been a genetic molecule, something like DNA or RNA, capable of passing along blueprints for making proteins, the workhorse molecules of life. But modern cells cannot copy DNA and RNA without the help of proteins. To make matters more vexing, none of these molecules can do their job without fatty lipids, which provide the membranes that cells need to hold their contents inside. And in yet another chicken-and-egg complication, the synthetization of lipids requires protein-based enzymes (encoded by genetic molecules)

It is easy to imagine the presence on earth of an enormous soup composed of a broad diversity of polymers that are intermingled and organized in more complex patterns such as RNA. Such pre-prokaryotic levels, of organization, without a cell wall, could evolve more rapidly and thus speed up the appearance of life. This would explain the implementation of the first three steps in Oparin's hypothesis.

The main problem of an open-ended creation of diverse interactions would have been a lack of controlled activity. Life could not have emerged until the appearance of an enclosed structure that contained the dynamic elements capable of producing the required mechanisms and processes. Scientists have utilized the properties of lipids to envision such an enclosed structure, the protocell. However, the protocell gets us only to the third step in Oparin's model. The fourth step, the emergence of life, remains unexplained. And it will remain unexplained until science accepts the essential need for the presence of a holism.

2. Abiogenesis—My View

If God exists, then the creation of the first living cell must have happened holonomically. God introduced telic configurations into the path of incrementation to successively create new forms of matter. At the descriptive level, the path of actualization would follow the steps in Oparin's hypothesis: formation of monomers; polymerization of complex molecules and formation of protocells. Given the premise, "God exists," the premise on which my thesis is based, then God was surely involved in abiogenesis, especially in the transition from a protocell to the self-replicating cell. The monomers, polymers and the protocells evolved according to the laws of chemistry and physics that are simply the manifestation, at the descriptive level, of the operation of the holonomic mechanism at the explicative level. In my view, the protocell, however, was the result of a specific telic configuration introduced, by God in the path of incrementation. The transition of the protocell to the animated cell occurred when the protocell encapsulated an immaterial substance. The molecular structure that we call a protocell which at the descriptive level was an enclosure of basic particles of matter in the form of lipids. At the explicative level, the lipid enclosure was a specific configuration of s-points that I refer to as a cellular s-frame. .

The animating substance that the lifeless protocell acquires to give it life is the same immaterial substance, the infinite nothingness, which activated the big bang. I call the infinite nothingness that is coterminous with the universe, *pneuma*. I call the pneuma encapsulated within the cellular s-frame, *bios*. Bios provides the means for the holonomic mechanism to operate at the cellular level and to animate the cell. The bios is either encapsulated within the cellular s-frame (the protocell) or simply associated with the s-frame's complexity. What is important is the idea that as the result of crossing a threshold of complexity, the bios effectively controls the cell and brings it to life. The observation that consciousness tracks complexity implies that the material (cellular s-frame) and the immaterial (bios) interact. It is through this interaction that a cellular algorithm controls the s-point configuration that represents the cellular s-frame, and we observe a living cell.

Chapter 15—Biological Stage

Once the earth had formed, it took the chemical elements, especially carbon, about a billion years of increasing material complexity before the third creation event, abiogenesis (creation of life), occurred. Abiogenesis introduced the biological stage with a significant increase in material complexity in a relatively short period of geological time with the creation of the first living cell after which followed a long period (*stasis*) of slowly rising complexity.

Abiogenesis is one of the great unsolved mysteries of science, and there have been hypotheses offered to close the explanatory gaps between the lifeless earth and the creation of a living cell. Most of the proposals address specific steps in the abiogenesis sequence without tying them all together in a single coherent process. The bottom line in all of this is that science cannot explain abiogenesis, the origin of life.

A. Cellular Life—Common View

My dictionary defines life as: "*The property of plants and animals which makes it possible for them to take in food, get energy from it, grow, adapt themselves to their surroundings, and reproduce their kind; it is the quality that distinguishes a living animal or plant from inorganic material or a dead organism.*"

I have two problems with this definition, first, it merely describes what we observe and does not explain what life is, and second, it misses the target in that, as explained above, life resides at the cellular level not at the animal or the plant level as implied in the definition. Science's explanation of life is that it is an emergent property based on an increasingly complex arrangement of organic molecules. Emergence is not an explanation that identifies the mysterious order of life that distinguishes animate from inanimate (or inorganic) matter; it is an abstraction that is just another way of defining life.

At least the dictionary offers a definition. Not so with biology books. Each book avoids the definition by presenting its own list of life's properties. For example, the following properties of life that appear in lists found in several biology books: complexity, organization, uniqueness, *holism, morphogenesis;* regulation, feeding, nutrition, excretion, *homeostasis,* development, unpredictability, growth, heredity, , evolution; mutation motion, movement, population, and irritability. Although every property of the list is associated with life, even the total list of properties taken as a whole does not suffice as an definition. Such a list can be reduced to three general catagories without which matter could not produce and sustain life; the categories are: *morphology, sustenance, and propagation.* Morphology appeared with the appearance of the first proto cell that contained neuronal information that provided sustenance and propagation. When present these requirements result in the most obvious observable property of matter: the thing we call life.

At the descriptive level, we observe life as change—especially the forms of change we call growth or motion. Although biology does not explain what life is and how it began, it is certain that life originated within a cell and cells remain the fount of all life. Animals and plants are not alive, their cells are. Plants and animals die because their internal mechanisms do not support the cells' life. Although the mechanisms required to support life are more complex in the multi-cellular organism than in the single-celled organism, the same three life- requirements are necessary; morphology, sustenance, and propagation apply to all forms of life. A living organism must support the life-requirements if it is to survive, hence the maintenance of morphology, sustenance, and propagation of its constituent cells affects the life of the individual multicellular organism.

1. Morphology—Cellular Structure

There are, in this world, too many kinds of cells to count, but they mostly fall into two basic cellular structures: (i) prokaryotic cells are simple cells without a nucleus; (ii) and the more complex eukaryotic cells that have a nucleus. Prokaryotes are the basic unit of single-cell bacteria; eukaryotes appear mostly in multi-cellular organisms, primarily plants, fungi, and animals. However, there are sufficient exceptions to the classification rules to give botanists fits when trying to categorize cells into a limited number of kingdoms.

Cells come in many sizes and shapes, but all share certain features in common. All cells are self-contained structures organized to provide an interior separated by a membrane-like surface that allows passage of materials containing the food that energizes the cell to pass in, as it allows the waste products of the energy-extracting chemical processes to pass out. The interior of the cell also acts as protection for the energy-producing apparatus and the *information* center that determines and activates each cell to perform its prescribed function.

The interior of the cell encloses a cytoplasm consisting of a translucent fluid called cytosol containing cellular elements such as mitochondria, plastids, and other organelles. Eukaryotic cells contain in the nucleus the DNA molecule containing the genetic information that governs the cells' life.

2. Sustenance—Metabolism

The mechanism called metabolism provides sustenance that means the cell must: (i) acquire food; (ii) transform it into energy (catabolism); (iii) construct new material (anabolism); and (iv) eliminate the waste products. Metabolism, the basic process by which cells function and grow, includes the conversion of food to energy and the use of energy to build materials and perform other biological functions that are important for growth and replication. Growth is important because it determines the timing of cell division. There is a wide range of cells in the biosphere, the size of which is determined by growth factors operating within the cell. Energy is the fuel that sustains life; life resides in the cell whose basic function is to use and supply energy. The cell is an energy-exchange structure. There are many ways that bacteria exchange energy, but most cells are either *autotrophs* or *heterotrophs*.

Autotrophs are organisms that absorb simple forms of energy such as light or simple chemicals that are converted and stored energy in complex organic molecules such as fats, carbohydrates, and proteins. There are two types: (i) *phototrophs* are autotrophs (plants and algae) that absorb light. Phototrophs using photosynthesis are responsible for producing and sustaining atmospheric oxygen that supplies most of the energy needed to sustain life on Earth. Photosynthesis uses plants absorb sunlight and carbon dioxide from the atmosphere and combine it with water to store glucose as energy, and to release oxygen to the atmosphere, and (ii) *chemotrophs* are those that absorb chemicals, and **heterotrophs** devised a more direct approach for acquiring energy, they consume autotrophs or other heterotrophs as an energy and material source in what might be an incipient food chain.

3. Propagation—Binary Fission

Prokaryotes came before eukaryotes hence developed the core processes to pass along to the eukaryotes the most important of which was the ability to propagate. All living things die; if any species, be it single celled or multicellular, is to survive, it must propagate. The first cell appeared on the earth equipped not only with many of the core processes such as metabolism and with information centers such as DNA, but more importantly, it possessed the ability to replicate, otherwise life would have been dead on arrival. Cells replicate by simple division. The prokaryotes replicate with a process called *binary fission*; the eukaryotes with a more complex process called *mitosis*.

One wonders if the initial appearance of life occurred with a single cell or with simultaneous appearance of many. A bigger question, however, is explaining replication. Certainly, the first cell or cells possessed a mechanism to grow and divide. DNA contains the information and how to divide, when to divide must have been a feedback mechanism of some sort.

B. Cellular Life—My View

The same holonomic mechanism, consisting of an input pattern, information, and an impetus that God uses to create and sustain objective reality can be adapted on a macroscopic scale to create and sustain cellular life. The telic configuration is a localized deviation from the main path of incrementation that is creating objective reality; a localized algorithm supplies information; impetus is the same cosmic impetus that increments the entire cosmic s-frame and all localized configurations. However, given the application of the holonomic mechanism on matter at the explicative level we can pose a definitive definition of life as: *the corpuspiration of self-controlled reconfiguration of a* cellular s-frame *(proto cell).*

1. Morphology—The Cellular s-Frame

By *"corpus"*, I mean for any living thing be it a single celled or multicellular, at the explicative level, is a specific, 3-dimensional configuration of s-points (a cellular s-frame) organized to form a membrane enclosing an inner volume *corpuspirated* with a particle of pneuma. I use the word corpuspirated, to describe this hylomorphic structure in which the physical cellular s-frame contains immaterial bios.

Note 23: I use the word "corpus" instead of 'body' to describe cells and multicellular organisms corpuspirated with an immaterial substance, bios, or nous. I use body when referring specifically to the somatic system of the material body.

A Corpuspirated cell is the fundamental structure for God's design of life. The encapsulation of a psychical particle (bios) in a closed and complex configuration of dedicated s-points (cellular s-frame) is manifested as a cell. The bios that is contained within the cellular s-frame contains a reduced version of the holonomic mechanism that functions to animate matter. Corpuspirated is how I describe the structure of the cell at the explicative level. It is the basic source of observed properties such as organization, uniqueness, holism, and complexity. It is a term you would not find in any other book of which I am aware. Corpuspiration is the mechanism by which self-organization of matter as a corpus occurs; it consists of a cellular s-frame and the encapsulated bios. A corpus is a well-defined structure that is unique to its kind, is whole, and complex. The corpus isolates and individualizes while at the same time confers a role as one of many. An interface—limited in its shape and size—isolates the interior of the corpus from the environment that allows corpuspiration and initiates life.

2. Sustenance—Cytonomic Mechanism

The structure of the cell, which at the descriptive level is a configuration of electrons and quarks, is at the explicative level, a configuration of s-points, a cellular s-frame. What we observe at the descriptive level as motion is, at the explicative level, life, the incremental reconfiguration of the s-points that form the cellular s-frame controlled by the bios from within using a specific form of the holonomic mechanism called the *cytonomic mechanism.*

The mechanism that animates the cell has the same form as the holonomic mechanism, that is, it consists of initial condition, algorithm, and impetus. The bios controls the change in configurations of the local s-points associated with the cellular s-frame in the same way that pneuma controls the incrementation of the cosmic-s-frame. The cell, of course, is also subject to the holonomic mechanism from without.

At the explicative level, a species-specific-cytonomic algorithm provides information that defines the sequential incrementation of the cellular s-frame that appears at descriptive level as cellular motion. Since it is the same substance as the pneuma, bios has available to it all the information contained in the infinite nothingness, the Mind of God, but with a diminished fidelity just as a piece of a hologram contains the entire image found on the complete hologram but with diminished clarity. This imparts contingency to living matter. Preprogrammed as a prototype determines the life of each type of cell. This results in the diversification of cell types, the manifestation of which at the descriptive level appears as evolution.

The algorithmic information realized from the pneuma appears in the structure of DNA within each cell. The bios determines the timing of the cellular events. The DNA determines what happens, the bios determines when it happens.

The cytonomic initial condition is a specific configuration of the s-points that represent the cellular s-frame. This initial condition manifests as its structure at a descriptive level. The cell structure then reacts to its environment as determined by a prototypical algorithm unique to each specific type of cell.

The impetus for the cytonomic mechanism is the Mind of God. Since the divine impetus increments every s-point in the universe, all s-frames increment simultaneously. What this means is that God sustains all motion in the universe increment by increment, the cytonomic mechanism merely supplies the information that determines the movement and activity of each cell. God through the application of the divine impetus sustains life.

Since Mind of God provides the impetus that increments every s-point in the cosmic s-frame, including those s-points corpuspirated as living matter, there is no individual impetus, all s-points increment simultaneously in the Divine flow. Cosmological times, in which things move at different paces, occurs because not every s-point changes position when incremented. *At the explicative level, life is simply the incrementation of the localized cellular s-frame that is given impetus by the Mind of God.* This is not a description; it is an explanation since it answers the question: *what is life?*

3. Propagation—Replication

We have no way of knowing whether the creation of life occurred once or many times, and there is the question posed in the last chapter as to whether it was a solo event or not?

Since there is no way of knowing—or ever likely knowing—according to the principles of science, the answer must be based on plausibility and that which makes the most sense. A solo event, whether it happened once or many times, given the conditions on earth 3.6 bya, seems to be as plausible as lighting a candle in a hurricane, in other words, implausible. The more likely scenario would be a simultaneous actualization of life in many protocells. This too would have challenged the laws of probability, unless God was directly involved, which means the most plausible scenario would be one in which God exists and created life in multiple protocells at the same time. Then holonomic mechanism proceeded without direct design input from God for most of the period during which evolutionary mechanisms diversified the cellular world.

C. Cellular Stasis Period—Common View

The cellular stasis period that followed abiogenesis prepared the way for the appearance of multicellular organisms, a morphic structure of diverse cells acting together with each type of cell providing a specific function. The appearance of the first multicellular organism represents an enormous increase in material complexity, an increase that was many orders of magnitude greater than the increase in complexity with abiogenesis. During a 3-billion-year stasis period, while the evolution of biological elements resulted in a diversity of cellular types, the symbiotic life processes of the biological elements, groups of diverse cells organized in symbiotic relationships of increasing complexity until with a relative suddenness multicellular life appeared in a creation event called somagenesis. When the first multicellular organisms appeared in the fossil record, they did so explosively. Thousands of organisms divided among 20+ phyla, including the first representatives of all modern metazoan phyla, appeared in the fossil record. Paleontologists call this discovery the Cambrian explosion.

> Note 24: I use the term multicellular to mean an organism in which a single complex enclosed body contains differentiated cells and the core processes that showed up in the Cambrian fauna.

Phylogenetic analysis is the study of evolutionary relationships based on genetic data such as DNA, rather than morphology, body types. It indicates that animals evolve from a single common ancestor. The germinal creature must have been simple, one of the Ediacaran, although it appears that they are not good candidates. Science has found no interim forms between single-cellular organisms and multicellularity. We can deduce what multicellular body-plans and components and what core processes appeared during this stasis period by examining the Cambrian fauna.

Cambrian fauna represents the foundation of somagenesis, although, because 20+ phyla appear in the Cambrian fauna, the origin of multicellularity certainly predated it. The Cambrian fauna, although simple compared to later forms, already possessed many if not all the body plans and multicellular core processes that produce the three prerequisites for life: morphology, sustenance, and propagation and their associated core processes: ontogeny, homeostasis, and reproduction. How and when these mechanisms appeared is a mystery, but it is certain that they did appear in some form before somagenesis could occur.

Based on chemical analysis of biogenic carbon, science determined that the first life appeared about 3.7 bya. Another finding that might have been biogenic carbon found in zircon crystals pushed the date back to 4.1 bya. There is no direct evidence, but when the first cells emerged, they must have been prokaryotes. They multiplied and diversified until the seas were teeming with life. However, more than unicellular diversification had to occur before multicellular organisms were ready to appear. Although no complexity gaps appeared in the POA chart during the cellular stasis period, there were new structural innovations and core processes that emerged with sudden moderate increases in complexity. Two of the more important structural innovations occurred when cyanobacteria and later eukaryotic cells emerged. Cyanobacteria appeared around a half-billion years after biogenesis and introduced photosynthesis which converts carbon dioxide into oxygen and created a life-sustaining atmosphere.

Around 1.4 bya, *endosymbiosis* created eukaryotic cells and evolved a nucleus. About 1.2 bya, reproduction, an essential core process required for the creation and propagation of multicellular organisms appeared. The several speculative theories describing the creation of reproduction are based on a common ancestor created as the result of the recombination of two unicellular organisms that share genetic material. There are no explanations as to how these recombined organisms end up as two genders that seek their opposites to recombine. Unlike the straight-forward replication of the body that unicellular organisms used to propagate, a multicellular organism uses a more complex multiple-step process that required several innovative core processes and mechanisms such as *germ cells, meiosis,* and *mitosis*. In addition, there was also the problem of explaining the creation of gender. Although the transition to the male/female would have been from an asexual organism. Asexual behavior appears among the species living today; in the animal kingdom, it appears primarily in invertebrates and insects; but seldom found in vertebrates; but it seems more plausible for gender to take a path through asexual beginning than for it to evolve directly. The appearance of earlier reproduction provided the basic requirement for propagation that allowed the jump to multicellularity to occur. The two other requirements—morphology and sustenance— arrived at the Cambrian period with proven core processes that produced the observed morphologies and provided for cellular sustenance. Ontogeny is the core process of morphology; homeostasis and metabolism are the core processes of sustenance; and reproduction is the core process of propagation.

Ontogeny is the structural core processes all animals use to construct a body; it consists of three mechanisms: cell differentiation, control of cell growth, and morphogenesis: (i) cell differentiation is the transformation of a single fertilized cell into a diversity of cells required to create the final organism. In the human, the simple *zygote* transforms into over 200 distinct cells; (ii) *control of cell growth* refers to the increase in the number of cells in the exact number required for the assigned body component; (iii) morphogenesis is the internal mechanism that determines the shape of the organism and the arrangement of its constituent parts.

Homeostasis fulfills the sustenance requirement for animals; it is the self-regulating mechanisms that maintain an optimal internal environment by controlling internal body temperature, tissue fluid, and the sodium, glucose, and water composition of the blood. Homeostasis must operate effectively at the cellular level, at the system level, and at the corpus level. That this complex system has developed mechanisms to address the needs across all three levels is amazing, and the body has somehow acquired ingenious solutions. Materialists claim the event occurred through stochastic selection of the fittest.

D. Cellular Stasis Period—My View

Whereas there are plenty of hypotheses and theories associated with abiogenesis there are only a scant few that purport to describe the emergence of animals, plants, and fungi.

Materialistic scientists find it difficult to apply evolutionary theory to the creation of sentient life. Whatever scant theories science produces to describe the creation of multicellular organisms, it will have to explain the development of ontogeny, the development of an egg into an adult, because the first fossils that arrived during the Cambrian explosion were *ontogenetic*. But there was more, much more that appeared in the path of actualization during the cellular stasis period. When complexity departed the abiogenesis event, it was with a simple prokaryote organism. When life arrived at the onset of somagenesis it carried in its baggage a multiplicity of morphologies, mechanisms, and processes that were prerequisites for somagenesis. There is no explanation as to when and why each of the actualities appeared. Furthermore, unlike abiogenesis, an all or nothing event, a thing is alive, or it is not, with somagenesis there are interim possibilities that we can imagine. My claim that life resides in the cells allows one to explain partial steps to multicellularity.

The 3-billion or so years from abiogenesis to somagenesis left no evidence of when or how each of the essential features of multicellularity appeared. The creation of life in this stasis period depended partially on the timing of events. My premise that God exists allows me to invoke the holonomic mechanism to explain the actualization of the core processes and material elements required for somagenesis to occur. The best I can do is to provide a general outline of a cellular automata type mechanism consisting of an initial configuration, information, and impetus, which can be, unlike mathematical equations, applied to describe biological observations. Because the algorithm is more versatile for explaining a diversity of phenomena, it provides the comprehensiveness that current scientific theories lack.

To complicate science's problem of finding fundamental explanations, a new phenomenon appeared with the actualization of multicellular life, the stuff we call consciousness. Consciousness, which is the manifestation of the psychical substance, has been increasing along with material complexity but would not have been noticeable until the multicellular *metazoans* evolved along the POA.

E. Somagenesis—Common View

No one knows for certain when somagenesis occurred. Defining multicellularity as any organism consisting of more than one cell allows us to set somagenesis back to 3.6 bya based on possible colonial arrangements or aggregations of simple unicellular organisms like *stromatolites*. This is a scientific sleight-of-hand intended to avoid an explanation of multicellularity associated with organisms as sophisticated as the Cambrian fauna. The hard evidence for the actualization of multicellularity is the body of fossils found in the Cambrian strata dated 540-490 mya when representation of most of the current phyla appeared. The fossils are all marine species arranged in 20–25 phyla. The fossil that closely represents our ancestor is called the Pikaia, a flattened wormlike creature that crawled along the sea floor. Yes, despite the common view taught in schools, it is not the apes we evolved from, it is a worm.

The suddenness with which the diversity and complexity of sentient life appeared in the Cambrian Period suggests that some form of somagenesis occurred at an earlier period. Trace fossils—impressions made by complex organisms called Ediacaran fauna—appeared in strata predating the Cambrian strata makes the supposition plausible. The Ediacaran trace fossils are multicellular organisms that look like modern-day things such as jellyfish, soft corals, and sea anemones. The Ediacaran appeared in strata dated from 630-540 mya after which they disappeared from the fossil record with no connection to the Cambrian fauna. Although the Ediacaran are not related to the Cambrian fauna, they suggest that the first multicellular organisms predated the Cambrian Period because the Cambrian fauna appeared with the body plans and core processes shared by all Metazoan. A more specific definition for multicellularity than "anything composed of more than one cell," would date somagenesis much later than the 3.6 bya dating. A more plausible date, for what I refer to as a seminal event, would be some time in the 600–700 mya period. The first multicellular organism appeared with complexity; orders of magnitude greater than the most complex single cell organism.

The Cambrian fauna, in which all the main components of life and representatives of the modern phyla appeared abruptly with surprising diversity without finding evidence of any transitional forms in periods leading up to the Cambrian. Since the Cambrian fauna appeared as 20+ species with complex morphology, it is apparent that, sometime before the Cambrian event, body plans and some forms of the core processes such as ontogeny, homeostasis, metabolism, and reproduction had evolved:

F. Somagenesis—My View

As the result of the sudden appearance of complete multicellular organisms and the tremendous increase in complexity associated with the Cambrian explosion, an explanation of somagenesis is one of the most puzzling problems confronting biological science. In any discussion of evolution, invariably the science community avoids discussion of creation of multicellular life; science does not even have a name for the multicellular creation event.

Somagenesis is an argument for the involvement of God in the POA; the theory of evolution does not explain somagenesis. Evolution addresses the modification of an existing morphology that results in the appearance of new species, while I hold that soma genesis actualizes innovative morphologies that had not previously existed. Science needs a hypothesis other than Darwinism to explain the creation of new forms of matter rather than the present explanation of new forms simply being modification of existing forms.

The final goal of evolution is the human species; the Cambrian fauna from which humans evolved may have been less complex but had the patterns for our current morphology and core processes. Although it is difficult to imagine the development of an organism that represents the last common ancestor (LCA) for the metazoan kingdom, it must have possessed the capability to reproduce ontogenetically from an ovum fertilized from a male gender to form a cell that then grows, differentiates, and produces not only a complete multicellular body, but one or the other gender. How does science explain ontogeny, fertilization, and gender arising out of a soup of bacterial cells?

When somagenesis occurred, it could have occurred more than once, and multiple species emerged concurrently as the multiple phyla of the Cambrian fauna seems to indicate. Since the explicative level is based on the existence of God and we invoke the "God did it" argument, here is how God may have done it: The formation of the corpus of the multicellular organisms (morphic s-frames) encapsulated another particle of the pneuma called *nous*. God provides the impetus to the morphic s-frame using a *morphic mechanism* to increment sequential configurations according to an algorithm just as the holonomic mechanism reconfigures the cosmic s-frame. Each cell has an exact copy of the DNA that contains all the information for implementing ontogeny. We can imagine how DNA instructions implement cell growth, but how could it implement the differentiation and morphogenesis processes without a blueprint to guide these processes?

Note 25: Just as a the cytonomic mechanism is the form of the holonomic mechanism adapted to the behavior of cells, the adaptation of the holonomic mechanism to the multicellular organism is called the morphonomic mechanism.

One possible way is that which the English biochemist Rupert Sheldrake (1942–) refers to in his book on morphic resonance as *formative causation*, a blueprint within a morphogenetic field. The mechanism called morphic resonance guides the formation of entities that had not previously existed. Somagenesis is the event in which the form of the first common ancestor was instantiated. I envision what Sheldrake calls the morphogenetic field is what I call the pneuma; and formative causation is the morphonomic mechanism; and morphic resonance is equivalent to what I call *noumenal resonance.*

God created the form of a common ancestor as a specific telic configuration of s-points that integrated the necessary core processes and material elements previously prepared during the cellular stasis period. The LCA was nothing more than an extraordinarily complex configuration of s-points, *a morphic s-frame* for which a single rational number on the real number line references the location of each s-point. The final configuration for a common ancestral morphic s-frame included nous that contained the basic algorithm for multicellular life. The result initiated the teeming biosphere formed by living entities, both single cells and multicellular. In addition to the appearance of consciousness, the elements that I show in figure 5 brought with them a new property with which the organism can contribute to the formation of each next s-frame; we call it *behavior.*

PART 3—Rational Reality

Chapter 16—Animal Behavior

Animals, along with plants and fungi, are the three main kingdoms of life. There are millions of species of animals that span a vast range of body types varying in complexity from sponges at the low end to humans at the high end. However, despite the wide range of body types, animals share certain properties such as: (i) eukaryotic cells and multicellular organization; (ii) consciousness; (iii) motility, the capability to move independently; (iv) heterotrophic sustenance—the ingestion of organic material; (v) reproduction; and (vi) aerobic respiration (breathing oxygen). Behavior of multicellular organisms simply means how a corpus (body) reacts to stimuli. Furthermore, the behavior I describe is that of a normal, well-adjusted person with normal functions and well operating faculties requiring no pharmacological assistance. We humans have at the core of our being an animal, but we differ from the rest of the animal kingdom because our range of behavior is immensely more complex. I simplify the complexity by reducing behavior to the four behavioral elements shown in figure 5. They are stimulus, corpus, motivation, and behavior. The corpus is the body of the animal; stimuli both external and

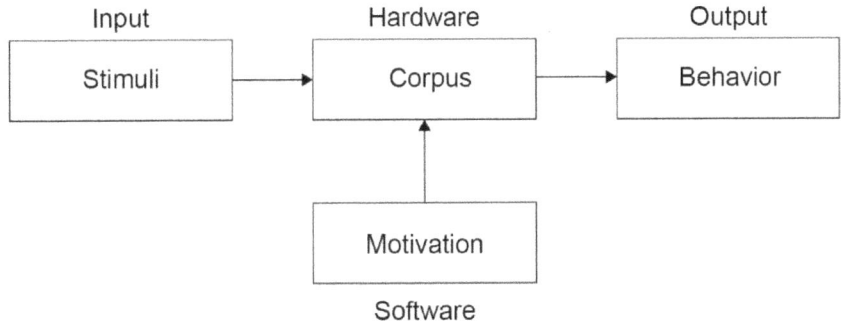

Figure 5. Behavioral Elements

internal generally tell the corpus when to act; motivation tells the corpus how to act and creates three levels of behavior: animal, rational, and spiritual. (i) Animal behavior is motivated by our basic instincts; (ii) rational behavior is motivated by tangential insight; and (iii) spiritual behavior is motivated by radial insight. My use of three motivational elements is not unlike Freudian psychology. For example: (i) my "basic instinct" mirrors Freud's "Id;" (ii) my "tangential insight" his "ego;" and (iii) my "radial insight," his "super-ego. The (i) basic instincts are: identity, security, and stimulation. (ii) Tangential insight includes intellect, reason, and values. (iii) Radial insight is based on the virtues of faith, hope, and charity. The behavioral elements acting individually or in combinations generate the following forms of behavior: reflexive, perceptual, procedural, rational, irrational, mental, and transcendental. The levels of behavior, based on the motivational elements are animal, rational, or spiritual, I identify stimuli as spontaneous or intentional and the main components of the corpus as material or immaterial memory.

Figure 5 shows the relationships of the four elements of multicellular life: stimuli, corpus, motivation, and behavior—that apply to all animals including humans. We humans are animals with the capability and, more often than is wise, the inclination to behave like animals. However, we possess a mind—a faculty that no other animal possesses—that provides the capability to behave at two other levels of behavior: rational and spiritual.

There are countless millions of species each with a specific mode of behavior. The range of animal behavior is as vast as the number of species. An animal's reaction to stimuli depends on the species to which it belongs, in other words, its basic nature. When hungry, herbivores graze, and carnivores hunt. All vertebrate behavior is based on similar core mechanisms. For example, all behavior results from the action of muscles. Whether we are running, eating, sleeping, smiling, speaking, or any other physical behavior, we vertebrate are activating the movement of muscles. Bioscience has thoroughly explored and described the muscular core mechanisms in mammals. The diversity of behavior of all non-human animals depends on the animal's place on the *cephalic spectrum* (CS).

> Note 26: The cephalic spectrum (CS) is the hierarchal rating of a species brain power as determined by the concentration of the nous in the brain—low for insects high for primates, a quantum jump of orders of magnitude for humans.

Insects behave reflexively. Non-human species on the high end of CS are, in general, capable of three behavioral modes, reflexive, procedural, and perceptual. Since we humans are animals, we also behave in the same three modes. Because of our ability to think, we humans are capable of additional modes, but at our most fundamental level of motivation, we are animals. We can observe and describe **how** and **when** animals act at a descriptive level. We cannot observe **why** animals act the way they do at any specific time; we can only explain why they act at the explicative level. We humans are animals when we act at an explicative level, and it is at the explicative level that we exhibit the basic animal nature that we hold in common with the rest of the metazoans. I include figure 6, a more detailed view of figure 5, to assist in the understanding of my view.

A. Animal Corpus

Scientists have described and cataloged close to a million of the estimated 8-million species. There are several ways the species differ, and although the variety of morphic s-frames is immense, there is a general shared blueprint consisting of the corpus, stimuli, motivation, and forms of behavior that form the pattern. We humans are animals and share a common behavioral pattern with the rest of the metazoa.

There is a single psychical substance—infinite nothingness—present in our world in three modalities: pneuma, bios, and nous. Pneuma operates objective reality; bios animates cells; and nous operates multicellular organisms. Nous contains the species-specific telic configuration and the algorithm to provide information for incrementing each morphic s-frame along the path of incrementation. What this means, at the explicative level, is simply, each species evolved a body plan that produced the form of behavior that best fit survival in a changed environment. For example, the creatures that swam in water gave up their fins in exchange for legs that produced the kind of behavior—running and breathing—that increased their chance of survival on land. Since God exists, God must have had a purpose for introducing legs. My purpose, at this point, is to direct the line of evolution towards the creation of humans. I present my answer to why God created humans in Part IV.

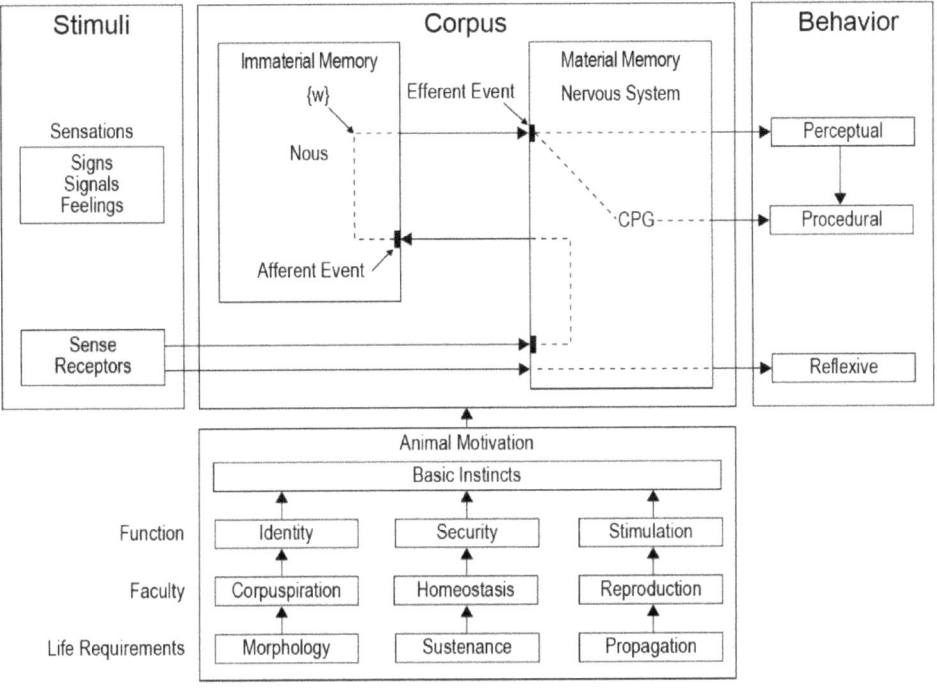

Figure 6. Animal Behavior

Somagenesis produced multicellular organisms consisting of bios containing cells immersed in the nous. The animal body represents a hylomorphic duality of material (s-points) and immaterial (nous); it also possesses—within the corpus—a psychical duality of the nous and the bios. Only animals are corpuspirated, plants and fungi are collections of specialized cells without an enclosed structure like the metazoan corpus that encapsulates nous. Since plants and fungi are not corpuspirated, they simply consist of collections of cells without nous, just bios—they are not conscious at a level beyond which the cells themselves provide.

Nous contains the power to interpose telic configurations associated with its own morphic s-frame; it determines its own path of actualization, or, in the more familiar terminology, nous determines the behavior of its own body. This is especially true in the human animal, for which behavior is associated with volition and the freedom to choose. We call the human capability to choose, *free will*, a property of the mind. Animals make choices that are not free; hence non-human animals do not have free will.

B. Animal Stimuli

Stimulation needs, arising out of the need to propagate, is essential to the survival of the species. Animals must reproduce. Reproduction translates into the basic need for stimulation, which translates into sexual encounters. The urge to mate in many non-human species is the focus for secondary forms of behavior such as territorial marking and defense, posturing and display, challenge and encounter and especially fighting. The human stimulation need that gives rise to sexual behavior experienced transference into other modes of behavior both: (i) positive behavior associated with survival such as hunting, exploration, adventure, games, and (ii) negative behavior such as fighting, battle, crime, addiction, and much irrational behavior.

1. Sensations

Animal stimuli consist of (i) external sensations such as loud noise or an unusual odor; and (ii) internal feelings such thirst, sudden pain, or sudden emotion such as anger. Because only humans have minds, the stimuli experienced by animals are solely physical which means the stimuli act through the nervous system, and no psychical element is involved. The stimuli that an animal experiences are signs, signals, and, or feelings. Because we have a mind, it also has the power to stimulate behavior. Stimuli determine when animals act. Sentience is the faculty that transforms sensations into qualia. But once stimulated into action, behavior can be reflexive, perceptual, or procedural. Most non-human animal behavior is spontaneous, i.e., behavior is initiated by sense receptors. Reactions are usually direct and sudden and involve material memory. The stimulus element determines when we act, its absence results in a state of repose.

Signs are external sensations that indicate a change in the environment that does not warrant immediate action. Signs are sensations that result in passive behavior. Signs are mostly amorphous. The brain, when conscious, is alert to all the senses and any one of them can lead to a change in behavior. A sign such as a sound, a smell, or an approaching storm causes heightened awareness and a change from rest to vigilance, sleep to wakefulness, or relaxation to alertness. But an animal does not act unless the sign carries a meaning and becomes a signal.

Signals are signs that have meaning; they are mostly sensations or physical feelings that have specific meaning and induce behavior, to act or not to act. The approach of a predator, or the sighting of prey induces the animal to flee, fight, hunt. A signal's itch, a pain in the leg, thirst, hunger are specific sensations with specific meaning that evoke action. In addition to sensations, signals can also arise.

2. Feelings

Subjective experiences that arouse the nervous system to fill cellular needs such as pain, hunger, thirst, discomfort, and arousal. Internal sense receptors in organs signal with the secretions from the endocrine system the need for air, food, water, sleep, comfort, safety, health, and any other need that threatens life if not met. In addition, there are psychological wants and needs which induce hormonal behavior, primarily emotional needs; they are present in humans and to a lesser extent in animals, especially animals domesticated as pets. Emotions such as fear, insult, and challenge can signal the person to act.

3. Sense Receptors

The specialized cells in the body that react to stimuli and excite impulses in the neurons of the peripheral nervous system (PNS). Sense receptors associated with external signs and signals act directly by sending impulses along sensory neurons to control centers in the material memory. Sense receptor associated with five main senses—sight, smell, taste, touch, hearing, are not the only receptors of sense that may initiate behavior. The body uses a variety of means to signal an internal need.

C. Motivation

The basic instincts consist of three requirements for the motivation of multicellular life: morphology, sustenance, and propagation. For the human animal, a corpuspirated morphic s-frame provides three functions: (i) morphology the source of the corpus (ii) homeostasis and metabolism the source of sustenance; and (iii) reproduction is the source of propagation. The three functions generate the following needs: identity, security, and stimulation.

1. Self-Identity

Self-identity is innate and primarily determined by our heritage, where and when, we were born including parents, nationality, language, place of birth, race, gender, intellect, phenotype (stature), physiognomy (appearance), cohort or generation, and class. We mostly acquire our collective identity; it includes occupation, avocation, political party, and many other tastes and preferences. We are born into certain acquired identities, such as our language, and our parents' political affiliation and religion. However, as we mature the forms of identity often change.

Our self-identity derives from the innate personal characteristics, both physical, mental, and from the *social units* we are born into. Hence, we acquire our gender, race, nationality, language, and religion (or no religion) by birth, but also acquire another set through our own choice. We can choose a profession, a team to root for, a political party, favorite pastimes, heroes, and other things with which to identify ourselves. We are most comfortable when among social units with which we identify; this provides social motivation, motivation that makes us feel most comfortable, safe, and confident of our behavior. If not modulated by insight the mob mentality leads to bullying, riots, gang wars and other more insidious forms of identity motivated behavior. It allows the truly evil leaders of large social units such as nations to garner the support of the followers whose identity derives from the social unit, to engage in wars of conquest.

Since every human's identity consists of multiple layers, each of which can express primary motivation. The priority of expression is inverse of the size of the social unit from which one's identity derives. At the lower level it is amazing how much irrational behavior is associated with political parties, fans of sport teams, religions, and races. The need for identity arises from morphology, a basic life requirement. Human morphology is corpuspirated; it consists of a body isolated from its environment that encloses nous, a psychical substance. The isolation of the corpus from its environment gives rise to an awareness of self, a sense of identity, the most powerful instinct motivating animal behavior. Our identity derives from two sources: a self-identity derived from heritage and a collective identity derived by choice.

We can emigrate to another nation; we are born into a specific religion but can join another; we can choose a political party; we can acquire a specific occupation, we can like certain music, types of books, types of food, art, and others. In other words, we can, to a certain extent, create our own image, our own personhood, our own acquired identity. In certain ways acquired identity determines the music we like, the clothes we wear, hair style, to acquire a tattoo, and many other elements of one's self-image.

We are born with a collective identity that includes family, community, and other civic organizations such as city, state, and country. We cannot easily change the status of our birth, our family for example, but about every other order of our collective identity is subject to change simply by moving. We can also acquire elements of our collective identity by joining social entities such as clubs, churches, teams, vocations, and avocations. Often collective identity is a more powerful motivation than self-identity.

Although a large part of our *collective identity* is determined by our origin it is usually not as strong a determinant as the identity that we choose for ourselves. As a motivational component of personal behavior, acquisitive identity has a stronger impact. Riots break out at sporting events simply because fans choose to identify with "their team" that is associated with their school or their town or completely arbitrary.

How a person behaves when the identity drive is engaged depends on that person's relationship to the social entity. A person can belong or become. By this I mean that when becoming part of a social entity a person can view it as a self-identity. We can belong to a social entity without it becoming a self-identity. Suppose you are a teacher, a comment like, "the trouble with teachers is," normally would not generate much of a negative response as the comment, "the trouble with YOU teachers is!" The first comment is a general statement of your collective identity; the second comment is personal. Although communication directed at self-identity tends to cause an individual to act, collective identity is the root cause of most of humanity's inherent problems. On the large scale it leads to war more often than any other cause. On a small scale it leads to bigotry, bullying and exclusion.

The second requirement for multicellular life, sustenance, generates a need for security. To be secure, we must eat, drink, avoid danger, and find protection from the elements. We humans find ourselves in situations where our survival is at risk and must resort to behavior that normally is irrational, for example, when soldiers are in combat, when firefighters enter a burning building to save a child. The conditions under which humans fight wars devolve into uneven degrees of rationality of the two sides of the affair. The same is true for first responders. Each situation determines the control individuals have over the need for security; in modern society for a large part of humanity, technology secures sustenance. Starvation occurs only in those parts of the world that are still susceptible to environmentally caused crop failure. However, personal consumption of food and drink generates pleasure that often seduces humans to bi-pass insight and overindulge in such consumption. This leads to personally harmful acts such as gluttony and greed, but these are minor irrationalities when compared to the consumption of alcohol and drugs. Buried in gluttony and greed there is at least a core of nutritional survival needs; alcohol and drugs have no positive value for the need for sustenance. Alcohol may have positive value when indulged in in small quantities; recreational drugs have no value whatsoever. Their consumption is completely irrational.

2. Security

Sustenance, the second requirement for life, translates into the basic need for security. The sustenance requirement depends on two internal mechanisms—metabolism and homeostasis—the mechanisms that provide sustenance. Metabolism is the inner process that converts food to energy and eliminates waste products. Metabolism means each cell in the body must: (i) acquire food; (ii) transform it into energy (catabolism) and (iii) construct new material (anabolism); and (iv) eliminate the waste products. To sustain metabolism, the animal must not only acquire food but also must maintain the proper conditions that allow metabolism to take place. Homeostasis is a mechanism that maintains the proper conditions shelter, and others,

The main motivation relative to meeting the need for security is the acquisition of food, for non-human animals in the wild, this is a continuing motivation. For humans, even when food is available, we must decide as to what, when, and sometimes where to eat. This is especially true in the modern world where we minimize starvation. In addition to the metabolic need to acquire food, homeostasis has its own needs. The body needs a narrow range of temperature, shelter, rest, avoidance of harm.

3. Stimulation

God created reproduction as behavior by which metazoans propagate. For most species, sex is the reproductive mechanism. The motivating factor is stimulation. It is the stimulation that induces multicellular organisms to propagate. Without the stimulation, what otherwise appears to be an irrational act would have come to a quick end and multicellularity would have disappeared. We humans, we thinkers, found that sex is not the only behavior that stimulates. Certainly, the predator species found stimulation in the hunt and the kill, and that kind of stimulation yielded a double reward: for sustenance and the stimulation. Stimulation, at times, derives from the identity need, especially self-identity. To rise in the hierarchy of the tribe required success in battle and in games of skill. The best hunter, fiercest warrior, the fastest, the strongest, the bravest all earned tribal esteem and the privileges that went with it. The combination of identity and stimulation became imbedded in the male animal genes that led to the most irrational behavior of all, fighting and full-scale battle and much of the benign irrationality of modern people such as mountain climbing, boxing, bull riding, and other stimulating activities. The last few examples have demonstrated the merging of the basic instincts in which more than one basic need is involved. In *The Territorial Imperative* by Robert Ardrey, he found that the acquisition and defense of territory filled all three of the basic needs of the animals he observed. They are identity, security and stimulation.

D. Behavior

An animal is a corpus reacting to its environment. The environment embraces the corpus in a sea of sensations, sights, sounds, odors, and feelings. We experience most without a reaction, but when the senses detect a signal, external sensations such as a noise, an odor, the approach of a predator, or the sighting of prey can either be a sign or a signal. Feelings are internal sensations, the change of which are signals. The usual state of an animal is one of repose.

In addition to examining the levels of behavior on the motivational elements, the main components of the corpus is material or immaterial memory. The behavioral elements acting individually or in combinations generate the following forms of behavior: reflexive, perceptual, procedural, rational, irrational, mental, and transcendent. I do not pretend to be knowledgeable in psychology and make no claim to the credibility of my view of human behavior, I merely offer, what follows, examples of how psychical objects such as consciousness, mind, and soul can be explained when we assume the existence of God and the hylomorphic nature of reality. Furthermore, the behavior I describe is that of a normal, well-adjusted person with normal functions and well operating faculties requiring no pharmacological assistance. The next three chapters deal respectively with animal, rational, and transcendental behavior. These elements—as animal, rational, or spiritual—I identify stimuli as spontaneous or intentional. I begin with animal behavior. Each species has developed its own modes of behavior for fulfilling the basic needs. The three needs are always part of an animal's motivation, with one or another dominating depending on which need requires the most attention. Although we, at the lowest level of motivation, are animals, our minds allow us to rise above the animal level of behavior when utilizing two more complex levels of motivation. I call such levels, discussed later, tangential insight, and radial insight.

E. Modes of Behavior

The range of animal behavior increases with cephalization. Animals on the low end of the cephalization scale always act in the same way in response to specific stimuli. The greater the cephalization the more complex the behavioral reaction. Although animal behavior is spontaneously physical, many species of animals (humans for example) are capable of intentionality. Non-human animals use only three lines of behavioral channels: two emerging from the sense receptors and one from the central pattern generator (CPG). The line that extends from the *will* to the CPG implies that some perceptual and most procedural behavior originates in the immaterial memory. Initiated by the senses, animal behavior is primarily spontaneous.

1. Reflexive Behavior

Reflexive is the simplest of behavioral modes; it occurs when the body spontaneously—without conscious thought reacts directly to stimuli through the sense receptors and efferent nerves in the PCN and CNS directly to the nerves of the muscle in the material memory The reaction may be very quick as when your hand touches something hot and you automatically withdraw it. The reflexive behavior of withdrawing one's hand from a hot surface is explained as the action of a hardwired network (motor memory) that starts with sense receptors at the end of a finger sending an impulse along an afferent nerve to the lower motor neurons in the spine where a synaptic connection is made with the efferent nerves that activate the appropriate muscles that moves arm and hand. This mode of reflexive behavior occurs quickly in short duration. In addition to physical reflexive behavior, humans often react psychically to sensations. Anger, fear, surprise, laughter, screams are among the most common psychical reflexive behaviors. They are all innate.

The Russian scientist Ivan Pavlov (1849–1936) discovered that dogs can be trained to salivate when exposed to different stimuli. It appears that animals on the higher end of the cephalization scale can learn reflexive behavior by training or conditioning. The amount of learning a species is capable of depends on its degree of perceptual intelligence. The human-animal is orders of magnitude higher on a learning scale than the next highest species. We learn reflexive behavior when learning to hit a baseball or stop at a red light.

2. Perceptual Behavior

Is behavior utilizing percepts and acting "without thinking" (without word images). Perceptual is the main mode of animal behavior; it depends on the processing of sensations detected in the sense receptors where action impulses are sent through nerves of the peripheral nervous system (shown in the chart as solid lines) to connections by afferent events with the central nervous system where the neural systems evaluate the percepts stored in the immaterial memory for recognition and chooses the appropriate behavior, then activates— through an efferent event— the muscles that initiate the behavior. Hence a leopard spies its prey, stalks, and strikes because of innate basic instincts. Although much behavior is instinctual, in most species, there are some animals for which minimal learning is necessary. For example, when young lions observe adults hunting, they are learning perceptually.

The most advanced animal behavior is learned-perceptual, when the sensation occurs in the immaterial memory before action occurs as when a prey animal flees a perceived predator. Perceptual behavior also occurs in humans, usually when the human is engaged in an activity that requires concentration such as when painting a picture or golfing. Perceptual behavior can be innate or learned. Without complex language, the highest mode of behavior that animals are capable of is learned-perceptual. It is based on the process of nonphysical recognition, meaning the young animal watches the adult demonstrate an activity and learns by perceptual memorization. The ability to memorize is innate. But the ratio of learned to innate increases with the cephalization scale; at the insect end, all is innate, and the amount of learned behavior maximizes on the chimpanzee end. Then, cephalization takes a significant increase with humans.

3. Procedural Behavior

Procedural behavior such as walking, running, chewing is routine, repetitive behavior produced by neural maps called central pattern generators (CPGs). CPGs produce rhythmic outputs without awareness. We humans utilize far more complex procedural behavior than animals do. Most animal procedural behavior is innate, but some is enhanced by observing what the adults do.

Chapter 17—Sentient Stage

The sentient stage began when the first complex multicellular organisms appeared, followed by the sentient stasis period of about 600-million years ending about 50-thousand years ago (kya) when the first human mind appeared. Other than trace fossils, multicellular organisms did not appear in the fossil record until about 540 mya when paleontologists discovered fossils in several locations in what is known as the Cambrian Explosion. Multicellularity represents an enormous increase in material complexity giving it the appearance of a creation event that I refer to as *somagenesis*.

A. Sentient Life—Common View

Although multicellularity includes animals, plants, and fungi that have numbered in millions of diverse forms, the objective of the path of actualization, based on the observation of increasing material complexity along with the associated consciousness, is the human being. Since the heart of the issue that I am addressing is the soul, it is the human line—the only organism with a soul—that I follow as sentient life. From the point of departure of somagenesis, the human line passes through animal, bilateral, vertebrate, mammal, placental, and primate stages and these are just the main stops, there are over 50 stops when counting inter stages.

Animals are complex, multicellular, eukaryotic organisms that possess self-induced mobility. All animals are heterotrophs, feeding on other organisms such as autotrophs. They possess differentiated tissues, a digestive chamber, an extracellular matrix in the form of shells and bones, and propagation with specialized reproductive cells.

If the basic morphology of life is the cell, then the morphology and processes of the multicellular organism exist to provide materials apart from the internal environment to sustain the life of each cell. We can understand multicellular organisms as a collection of diverse cells with specific properties and functions by considering the same three requirements for cellular life, namely: morphology, sustenance, and propagation. Each one of these requirements has associated with it a specific core process: i) ontogeny sustains morphology; (ii) homeostasis and metabolism sustain sustenance; and (iii) reproduction sustains the propagation of the species.

> Note 26: The following discussion of morphology, sustenance, and propagation applies to humans, not various transitional species. The text is impossible to understand without illustrations, especially the discussion on morphology. The subject is included to illustrate the complexity of life.

These mechanisms had to have appeared simultaneously. Ontogeny, homeostasis, and reproduction are interrelated, and they could not subsist as separate mechanisms. Cell differentiation showed up in the fossil record around 2 bya. However, cell differentiation alone is not evidence of multicellularity in the sense implied by a common ancestor.

The Cambrian Explosion presents challenging and important questions because it represents the time during which the main branches of the tree of animal life became established. It does not create a challenge to the fundamental correctness of the central observation of evolution—that all living species descended from a common ancestor. The pre-Cambrian period in the history of life extending over millions of years, provided time for the evolution of new body plans to have occurred before the Cambrian era. However, science does not have a description of how somagenesis occurred. This is understandable if we consider that there is not even a good explanation for abiogenesis and any scientific hypothesis for explaining somagenesis would be orders of magnitude greater. To understand how much greater the complexity of multicellularity is, we can explore a simplified view of sentient life and its three main requirements: morphology, sustenance, and propagation.

1. Morphology—Ontogeny

Science has identified a million of the estimated 8-million metazoan species. To discuss the nature of the animal requires us to find common factors among the millions of variations, i.e., to imagine a generic corpus. The different metazoan body plans are asymmetrical, spherical, cylindrical, radial, and bilateral. The human body plan falls into the bilateral category. *Ontogeny is the* core process involved in the formation of all the body plans of multicellular morphology. Ontogeny consists of three mechanisms: cell differentiation, control of cellular growth and morphogenesis, the development of the body's shape and morphology.

Cell differentiation is the mechanism by which one type of cell transforms into an entirely different type. This involves changing from a universal type (a stem cell) into a specialized type (a neuron for example). Differentiation occurs by binary fission when the parent cell divides into two cells, one of which is another stem cell plus a daughter cell that is a different type. The daughter cell has a different shape, size, and functions despite having the same DNA sequence. Biologists have described the differentiation in detail through a process called *gene expression.*

Note 27: Gene expression refers to the transcription and transfer of a protein. When a specific protein is made, we can say that a specific gene has been expressed. A gene is expressed when it is "triggered" by a specific factor.

Control of cell growth involves the amazing feat of transforming a single cell into a trillion cells simply by division. Biologists can describe exactly how that happens, but the control of cell growth and cell differentiation is another matter. The problem is to create a specific number of each type of cell by cellular differentiation then to place through *morphogenesis* each cell type in the exact position relative to all other types.

Morphogenesis is the mechanism of morphology that develops the body's shape and forms the body. There are countless millions of body types among the multicellular species of life that materialistic biologists claim evolved from a single body type using the same mechanism of growth it is called *embryogenesis*. It is also possible to imagine cellular differentiation as an effect of hormones on stem cells, but it is truly overwhelming to imagine the feat of generating the correct numbers of each type of cell, produced at the exact time, and placed in its exact position. And are we to believe that this takes place in the creation of billions of humans each decade with a success rate in the high 90% solely based on a stochastic mechanism called *gene expression*? I do not think so! Gene expression merely moves the mystery to another level without explaining how each stem cell knows what to become and where to go, especially since every single cell in the body has the exact same information stored in its DNA. And when all is in place and the finished product is delivered, the finished product subsists through an amazing set of internal processes designed to meet all the requirements of keeping the cells alive. Morphogenesis begins with the fertilization of a female egg by the male sperm. The egg and the sperm are haploid cells (each containing half the chromosomes) called gametes that combine to form a single diploid cell (contains a full complement of chromosomes) called the zygote consisting of a plasma membrane surrounded by a thicker coating called the zone pellucida. Figure 7 is a schematic representation of the cellular paths of human morphogenesis. It is presented here as an aid for discussion. The chart cannot begin to exhibit the complexity of the real process. The first four stages lasting about nine weeks represent embryogenesis, the development of the embryo. The last two stages represent the development of the fetus. The chart is a simplification of an extremely complex process. It is meant merely to outline the complexity involved in the formation of the body.

Blastulation begins as stage 1 when the zygote divides itself 5 times without growth to produce 32 smaller cells called the *morula*. In blastulation, the 32 cells multiply and differentiate to form a *blastocyst* composed of a grouping of cells called the *embryoblasts* that are contained in an outer layer called the *trophoblast*. The cells of the embryoblast, the *inner cell mass*, compact to one side of the trophoblast leaving a cavity on the other side called the *blastocoel*. The inner cell mass differentiates, and forms two layers called the *bilaminar embryonic disk* consisting of the *epiblast* that will form the embryo and the *hypoblast* that forms the umbilical vessel.

Gastrulation begins when the bilaminar embryo begins to elongate into the shape of an asymmetrical egg-shaped oval in which a groove forms in the central axis at the narrow end. The groove, known as the *primitive streak*, is composed of multipotent stem cells that migrate through the groove to: (i) form the endoderm layer replacing the trophoblast cells; and (ii) form a mesoderm layer in the middle. The resulting structure representing stage 2 is called the trilaminar embryonic disk; it consists of the ectoderm, mesoderm, and endoderm. Cells from the mesoderm form a rod-like structure called the notochord that provides a signaling and development function. This ends stage 1 and 2.

Neurulation, the process represented by the parts of the embryo depicted in stages 3 and 4, coverts the flat trilaminar elongated oval shaped disk formed by the three germ layers, through cellular growth, differentiation, segmentation, involution, and folding into tubular shapes.

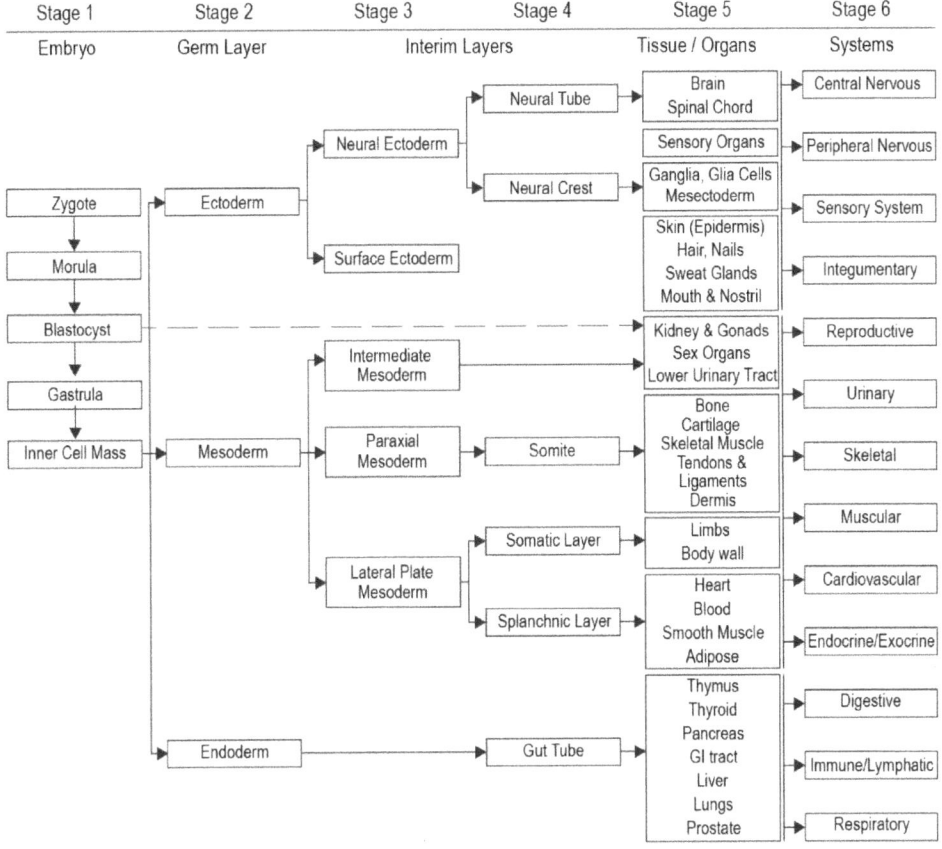

Figure 7. Morphological Elements

The ectoderm groove folds down into the mesoderm to form the neural ectoderm just above the notochord. As the part of neural ectoderm forms the *neural tube* it separates from the lateral sections that close to form the surface ectoderm. Meanwhile further differentiation creates the *neural crest* above the neural tube and the folding of the surface ectoderm downward enfolds the rest of the elements to become the outer layer (skin) of the embryo.

The mesoderm segments into the paraxial mesoderm in the center and the intermediate mesoderm on both sides and the lateral plate mesoderm on both outer edges. Continued signaling causes the later plate to split into a somatic layer and a splanchnic layer.

The endoderm forms into the gut tube when both sides of the splanchnic lateral layer folds downward taking the lateral part of the endoderm layer with it to form a tube with the mesoderm on the outside and the endoderm on the inside. Finally, the two laterals of the somatic layer fold downward and the two ends meet and form the body cavity. When completed a top view of neurulation reveals the body cavity enclosing the neural tube and the gut tube. Simultaneously the embryo was folding in the other direction to position head on one end and the spinal column on the other.

Organogenesis (Stage 5). Embryogenesis differentiates and organizes the proper cell types in their assigned position and produces four types of tissues—epithelial, muscle, connective, and nervous—that are the building material of the 70+ organs in the human body. An organ is any bodily structure formed from at least two kinds of tissue working together to achieve a specific purpose. The chart merely shows a simplification of the complexity of the cellular paths depicting the transition from the interim layers to the tissue and organ formation as indicated in the chart.

We cannot pictorially depict the complexity that is associated with the interaction of the four types of tissue. For example, we find the three types of muscle tissue—skeletal, smooth, and cardiac—in many of the organs. Placement of the tissues and organs in locations positioned as part of the 10^{+13} (depending how they are categorized) systems in the human body. All of this must occur with very precise timing. This, of course, is a growth process in which things must happen in the proper sequence.

The chart depicts the current systems of the human body that took 540-million years to evolve from the primitive wormlike sea creatures such as the Pikaia found in the Cambrian fossils. Given the time span of 540 million years, it is not hard to imagine the transition from Pikaia to human happening through the process of evolution. Pikaia possessed primitive versions of muscles, body cavities (coelom), closed circulatory system, and a nerve cord formed from an embryonic germ layer. Although we might imagine the evolutionary transition happening in the time since the Cambrian era, we would only be pushing the embryogenesis mechanism back in time. This implies that the unanswered questions associated with how embryology occurred were present from the beginning of multicellularity that obviously predated the Cambrian fauna.

2. Sustenance—Homeostasis

The life requirement that sustains life of the cells in a multicellular organism is homeostasis, the group of mechanisms that sustains metabolism. Metabolism is the cell's internal mechanism that allows it to function, i.e., to live. Multicellular metabolism consists of the acquisition of food, its digestion, and removal. Homeostasis refers to the self-regulating mechanisms that provide the optimal internal environment for the body and the optimum external environment that allows the cells to survive. Homeostasis must achieve its objective of providing an optimal internal environment by compensating for variance in: (i) the external environment; (ii) the input of food and drink; (iii) and the activity of the animal when both resting or working. What the cells need to live is a narrow range of temperature, a specific content of the bodily fluids, and a constant supply of essential chemicals and nutrients for metabolism to sustain cellular life. This means that homeostasis must regulate within an acceptable range, chemical and physical factors such as: the core internal temperature; concentrations of sodium, calcium; and oxygen, the glucose in the blood; the *ph* of the extracellular fluid, the volume of body water; and blood pressure. A separate complex regulating feedback mechanism monitors and adjusts each one of these homeostatic requirements. When a required factor increases or decreases, the body has mechanisms that compensate to bring the factor back to the optimum. For example: sweating compensates for high body temperature by dissipating heat, and shivering does the opposite by creating heat to compensate for low body temperature. Homeostasis is one of the core processes that had to appear with the first multicellular organism although in a far simpler form than present in humans.

3. Propagation—Reproduction

Whereas morphology and sustenance are associated with survival of the individual animal, propagation is associated with survival of the species. As a result, it is a special case in that the mechanisms and genetic code transfer from parents to offspring. Unicellular life propagates by simple division; sentient life propagates by a far more complex recombination mechanism. Unicellular propagation is an analytic process in which the whole divides into parts. Multicellular propagation is a synthetic process in which two entities enjoy forming a single new entity. Unicellular propagation requires a single mechanism (binary fission) to replicate. Sentient life requires gender, fertilization, binary fission, embryogenesis, meiosis, gestation, and nurturing. Describing how abiogenesis occurred is far simpler than describing somagenesis, and no one comes close to explaining either one.

The multicellular reproductive process begins in the early stages of embryogenesis when the gender mechanism commences during gastrulation and cells of the inner cell mass, separate into the somatic cells that form the body and the primordial cells that migrate to the insipient gonads where they form separate gender gametes. I show this special path, in figure 5, as a dotted line originating during gastrulation. This is another core process that had to be inherent in the first multicellular organism. With a stretch of the imagination, I can mentally devise a group of diverse cells gathered with each type accomplishing a different function, but how will it develop an epithelium enclosure to form a specific body plan and at the same time develop gender, a female and male, two different morphologies in the same species?

B. Sentient Life—My View

There is no theory that explains how ontogeny occurs. The prevailing view is that DNA and RNA use a mechanism called gene expression to drive ontogeny. This certainly must be the case, but the question then becomes how does DNA do this? It certainly must involve a complex mechanism. How can the information contained in the DNA, within the interior of a stem cell, instruct a cell to become a neuron of a certain size and to move to a place in the nerve that monitors temperature at the fingertip of a fetus? Since each cell in the body has an identical copy of the DNA, i.e., has identical information, what is the mechanism that differentiates and guides the cell to its proper place in the body? No one knows, and none of the mechanisms found in science's current knowledge, especially the theory of evolution, offers an explanation.

We can never be sure that science will not find a complete materialistic explanation of ontogeny, but that alone will not replace "God exists" with, "God does not exist." I contend that God will always create and sustain at the explicative level at which ontogeny along with all other new phenomena depend on God's introduction of telic configurations into the cosmic algorithm that then manifests at the descriptive level as new morphologies and the new phenomena to sustain them.

Now let me point out that there is a single immaterial substance, infinite nothingness, present in our world in three modalities: pneuma, bios, and nous. Pneuma actualizes objective reality; bios vivifies cells; nous animates multicellular organisms. As pointed out earlier, all life resides in cells; then God designed the multicellular corpus to maintain cellular life, consequently and in a sense has a life of its own whose requirements are the same as the cellular life requirements, namely, structure, sustenance, and propagation. Each member of every species, whether it is a single cell, an ant, a mouse, an elephant, a shark, an eagle, or a human, is a morphic s-fame with a specifically structured body. Nous controls both the specific mechanisms that maintain cellular life, and the process for reproducing. In addition to the morphonomic mechanism that induces behavior, nous also possesses vibrant properties that are manifested as subjective experiences such as qualia.

1. Morphology—Morphic S-Frame

An important question: when is a body corpuspirated, i.e., filled with the nous? To maintain consistency with the claim that the holonomic mechanism is the architect for all actualities, the nous must have been corpuspirated at fertilization and before the process differentiation of the cells. The holonomic mechanism implements the differentiation and is guided by a pattern (noumenon) that when applied to a multicellular organism, creates a morphic s-frame.

The *morphonomic mechanism* must be present for morphogenesis to take place. In the fertilization process, cellular s-frames (ovum and sperm) combine and share not only their *genome* but also share their bios that combines to form the nous of the multicellular corpus. Since nous is the substance of the soul, an incipient soul corpuspirates the zygote at fertilization. And here I am walking on the thin ice of heresy, since the subject is veering into the discussion of the creation of the soul and the choice of doctrines addressing the subject. The path one takes requires a choice of two doctrines: generation or creationism. There are two doctrines in the generation branch: (i) *traducianism* means the infant soul is generated organically from the souls of the parents, and (ii) when the term generation is used directly it means that the transmission of the soul from the parent soul in some unknown way. The doctrine of creationism holds that God creates every individual.

2. Sustenance—Morphonomic Mechanism

At the explicative level, a morphic s-frame, the configuration of s-points that defines a corpuspirated zygote encapsulates the nous that contains the morphonomic mechanism . Consequently, the holonomic mechanism determines the behavior of the cosmic s-frame and the cytonomic mechanism determines the behavior of cells, the morphonomic mechanism determines the behavior of multicellular organisms. The telic configuration in the morphonomic case is the configuration of s-points that initiates the formation of a zygote. In all animals the zygote is corpuspirated with nous at conception. Cell division begins with the morphonomic mechanism within the cavity of the zygote. There is an algorithm for each species of bacteria, (soma-cell) and multicellular organism that defines each prototype. Just as the algorithms for the prototypical bacterial and soma cells are instantiated as c-noumena in the pneuma of the realm of potentiality, the algorithm for each multicellular organism transfers from the pneuma to the nous. The nous contains the species-specific telic configuration (formative cause) and the algorithm to provide information for incrementing the morphic s-frame through the path of incrementation.

So far, I have introduced three modalities of the same mechanism, or the form or the blueprint of such, namely a cellular automaton form consisting of: (i) initial (telic) configuration, (ii) algorithmic information, and (iii) divine impetus. The initial configuration is flexible, the information is prototypical, and the impetus is uniformly constant. It is the same mechanism with which God created the universe, it is the holonomic mechanism. The holonomic mechanism created the Earth leading to abiogenesis; the cytonomic mechanism created the biosphere leading to somagenesis; and the *morphonomic mechanism* created Homo sapiens leading to the psychogenesis and the actualization of the mind. The holonomic mechanism is telic and a top-down function that produces growth, *motion, and movement.*

Gene expression, the scientific explanation of ontogeny is a bottom-up process; morphonomic mechanism is top down in which the psychical modulates the physical by guiding the process to produce a specific objective, hence it is telic. For animals, the morphonomic mechanism applied to all three functional changes—growth, motion, and movement—guides multicellular behavior. The mechanism creates the corpus (form) of the animal applies to the growth function at fertilization as it modulates the gene expression through cell differentiation, control of cell growth and morphogenesis, and the functions. Biologists describe every step of embryogenesis, namely the mechanisms, molecules, and processes involved in ontogeny, but science cannot tell us how certain embryological objects are determined. For example, biologists invoke gene expression to describe how cell division leads to creating growth but cannot describe the creation and precise placement of a specific number of cells of differentiated types to form each organ, bone, and muscle. Gene expression and cell division is a bottom-up process, the precise number of cells, a limiting factor to the process, must be top down. We can assume that the intrusion of a limiting factor such as enzymes, the environment, or other explanations for how a specific number of cells are formed, but then the problem becomes one of timing and the question becomes: what initiated the factor? And then what initiated the initiating factor? The easiest explanation is, of course, God did it, with an information stored in the realm of potentiality and transferred to the nous in the realm of reality. So that is how morphogenesis proceeds physically.

3. Propagation—Morphic Resonance

Just as abiogenesis could occur, required binary fission or its equivalent before some form of reproduction had to exist before somagenesis could take place. Since reproduction is essential for creation of multicellularity, to say that somagenesis could take place apart from a reproductive process is to put the chicken before the egg. The best solution is to put the egg before the chicken, and to have a pre-existing (noumenon) form of a chicken. An asexual process from which natural selection could create gender. No matter how somagenesis occurred we can be certain that it was not the result of the natural selection process; it had to be a morphogenetic mechanism. And for me it had to be a creative event.

C. Sentient Stasis Period—Common View

After somagenesis and the creation of sentient life, the sentient stasis period began as multicellular organisms diversified and formed three main multicellular kingdoms: fungi, plants, and animals. I concentrate on the mammal class of the animal (metazoan) kingdom with a concentration on the human animal; humans are the only metazoan associated with the next stage, the creation of the mind.

The graph in figure 3 shows a greater slope for the sentient than for previous stasis periods. I contend that: (i) because of increases and decreases in the number of species, there is, on average, moderate increase; (ii) because of the increase in complexity of available body plans there is an enhanced increase; and (iii) because of the time elapsed in which the increases occurred is shorter than for previous stasis periods. The sentient stasis period is the age of fossils. Prior to somagenesis there were only trace fossils and chemical analysis to aid in describing the history of the earth. Fossils appeared suddenly in the

Cambrian Period and have been sufficiently plentiful to produce a sequential history of sentient life.

1.The Geological Path

Geologists tracing the geological path through the layers of strata provided a vertical record that allowed them to create a record of biological evolution. The vertical arrangement of the Earth's strata, called the geologic column, provides the chronology of the fossil record. There are two general levels: the Precambrian eon and the Phanerozoic eon that has three eras: the Paleozoic, Mesozoic, and the Cenozoic. The Precambrian eon is associated primarily with the geological path. The Phanerozoic eon tracks the sentient stage and is associated primarily with a biological path as shown in figure 8. It took 500 million years for the earth's surface to stabilize, and there are still earthquakes and volcanoes reforming the earth's surface. Throughout most of this sentient stasis period the surface of the earth was still changing through a process called continental drift. Seven large *plates* and many small ones moved around and collided with one another forming mountains, volcanoes, and earthquakes. There were also large swings of the Earth's temperature that resulted in the extinction of entire species. There were five major extinction events during this stasis period that killed off a sizable percentage of the species. In general, extinctions occurred with dramatic changes in the chemical makeup of the atmosphere and the oceans brought about by combinations of several different causes such as continental drift, meteor impact, and volcanic action. The main extinction events defined the various geologic periods as follows:

The Ordovician-Silurian event (O–S) occurred 450–440 mya when 60–70% of the species went extinct. Due to the formation of the Appalachian Mountains falling sea levels led to vast quantities of plants that removed too much carbon dioxide from the air resulting in a global drop in temperatures resulting in increased glaciation

The Late Devonian-Carboniferous event (D–C) occurred 375–360 mya when 70% of the species went extinct. Giant land plants released excessive nutrients into the oceans causing mass blooms of algae depleting the atmospheric oxygen. Ash from volcano eruptions caused land temperatures to drop leading to extinction of land animals.

The Permian-Triassic event (P–Tr) occurred 250 mya when 96% of the species went extinct. The cause was an enormous volcanic eruption filling the atmosphere with carbon dioxide that resulted in acidic oceans.

The Triassic-Jurassic event (Tr–J) occurred 200 mya when 90–96% of the species went extinct because of an asteroid impact, climate change, and flood basalt eruptions.

The Cretaceous-Paleogene event (K–Pg) occurred 66 mya when 75% of the species went extinct including the dinosaurs as the result of climate change caused by volcanic eruptions plus steroid impact.

Extinction of multiple species opened environmental niches for new species to arise, for example had it not been for the K–Pg event that eliminated the dinosaurs, the mammals may not have had a chance to thrive. Also, it was lower temperature ice ages that caused all the extinction events, not global warming.

Periods	Mya	Plant Path	Biological Path
		Flora	Fauna
Cambrian			Cambrian fauna
	510		Invertebrates—trilobites
Ordovician	485		Glaciation
	444	Roots	Warming—jawed fish
Silurian	419	Ferns/seeds	Fish explosion
	400	Land plants	Land animals, wingless insects
Devonian	375	Fungi	Amphibians
	370	Primitive trees	
	325		Mammal ancestor (synapsids)
Carboniferous	320	Coal forests	Insects,
	299		Amniotes/ reptiles (diapsids)
Permian	275		
	250	Conifers	Dinosaurs,
Triassic	201		
Jurassic	200		Birds, mammals
	145	Flowers	
Cretaceous	66		Dinosaur extinction
	66		
Paleogene	34		
	23		
Neogene	3.0		Extinct. large mammals;
	2.6		Hominids
Quaternary	50 ka		Homo sapiens

Figure 8. Phanerozoic Eon

The observation that the universe is expanding is one of the two most important discoveries leading to science's description of the path of actualization; the second is the fossil record. The observation of an expanding universe led to the big bang theory; the fossil record led to Darwin's theory of evolution. We have already discussed the expanding universe and its description of the creation of objective reality, now we can discuss the fossil record.

2. The Fossil Record

Whenever humans dug into the earth, they often discover fossils. However, it was not until the 18th century that a greater interest in fossils led to the development of paleontology, the science of fossils. Concomitant with paleontology, and interdependent with it was the development of geology, the science of the earth's structure. The surface of the earth consists of layers (strata) formed by sedimentation then altered by the impact of wind, water, and other natural processes. It is reasonable to assume that a stratum closer to the surface represents more recent activity than a stratum at deeper levels. Scientists have developed methods to determine the age of rocks; hence the layers of strata provide a geologic calendar that records with relative accuracy the periods of Earth's history. Scientists name each stratum and group them sequentially as eons, eras, and periods. There is further division of periods into epochs and epochs into ages. Shown in figure 8 is the Phanerozoic eon. A fossil's age is determined by the strata in which it is found. The natural geologic processes plus the abundance of life on Earth over the ages has resulted in an abundant fossil record. When considered in respect to the geologic time scale, the fossil record presents a natural progression of increasingly complex life; the fossil record is to paleontology what the expanding universe is to the big bang. There is also a geological path in time that is associated with the history of the Earth's surface and the other factors that impacted life. Together with the biological path, the geological path has generated a wealth of fossil data with which to construct a history of the Earth and the living things it contained.

3. The Theory of Evolution

The basic aim of all living organisms is to propagate the species. The need to propagate manifests itself in the way that life has structured its elements to allow each species to sustain itself and to propagate its own kind within a specific environment. Then, if the environment changes in such a way as to lessen the ability of a species to propagate in a sustainable way, one of three things can happen: (i) the species undergoes genetic drift; (ii) a new species forms; or (iii) the species goes extinct.

In 1859, Charles Darwin proposed a theory to explain the apparent gradual change found in the fossil record of organisms from primitive to more advanced forms. It became known as the Theory of Evolution, and it led to the ongoing controversy between science and religion. Darwin proposed that species evolve through a process known as natural selection. Certain individuals of a group are born with a chance variation in the species characteristics. Sometimes an individual born with such a variation proved to be more adaptable to their environment and had a better chance of survival. Such individuals stood a better chance of producing offspring and thus perpetuated the chance variation in their progeny. Hence, a species acquired new characteristics over long periods of time even to the point where a new species emerged.

The mechanism that Darwin proposed for the introduction of change, chance variations, did not explain, at the time, the changes observed in the fossil record. However, near the end of the nineteenth century work done by the Austrian priest/biologist *Gregor Mendel* (1822–1884), was rediscovered. His work demonstrated that it is genes—some of which persist through more generations than originally thought—that pass on morphological characteristics. At about the same time several botanists discovered that spontaneous large variations, mutations, can occur in the gene pool of a species. This put Darwin's theory of natural selection on a firmer footing and became the accepted explanation for the fossil record.

When combined with the understanding of the role of genes in carrying characteristics from generation to generation, Darwin's theory became Neo-Darwinism, the theory taught in the schools of the world and is the thing people normally refer to when they talk about evolution in the popular media. Neo-Darwinism is an example of literal abstraction presented as a sequence of factors. The factors that are necessary for speciation to occur are:

a. Mutation in the genes that changes the characteristics of the organism.
b. Exposure of a population to change in the living conditions either by isolation, migration, and or environmental change.
c. Increase survival for the organism possessing the mutant characteristic.
d. Increase in the mutant organism's opportunity or chance to propagate.
e. Transfer of the mutant gene to the progeny. Sequential repetition of the first five steps over an extended period.

Extrapolating the evolutionary tree back in time leads scientists to conclude: (i) that life arose spontaneously from inorganic molecules; (ii) that physical mechanisms explain life and all its emergent properties; and (iii) that life reduces to the action of the fundamental physical forces operating on the simplest forms of matter. This is the mechanistic or reductionist view of creation. Assuming only matter exists, materialists contend that life arose solely from the action of inert atoms.

The main objection to evolutionary theory is not that it is anti-religious, but that conclusions go beyond what the facts support. Whereas the theory of evolution is about gradual change in phenotype over prolonged periods, Neo-Darwinism is the extrapolation of the theory of evolution to explain phenomena not observed in the fossil record, namely the contention that humans can trace their origin in small incremental changes over extended periods to an accidental combination of inorganic chemicals. This denies the existence of a psychical agency and hence denies the existence of God. That is a colossal conclusion to derive from the fossil record. However, by weighing the Neo-Darwinian theory of evolution against the creationists' literal translation of the Bible, materialists gain the edge in the intellectual debate. Once Darwinism was established as a tenet of the modern paradigm, the materialists used evolution as an argument to implicitly extrapolate to the beginning of creation as we have seen with cosmologists' attempts to explain a reality that has no beginning.

Evolutionary theory does not answer all the questions the fossil record raises, and there have been modifications throughout the 20th century; to become "the modern synthesis." Nevertheless, evolution is a more plausible explanation than its main challenger, creationism. Creationism is based on the literal translation of the bible which indicates the life span of the human species as 5 – 6000 years. This flies in the face of the facts found in the fossil record. In consideration of the observations made by geologists, paleontologists, and others, Neo-Darwinism is a more reasonable description of the fossil record than that given by creationists. In fact, those on the evolution side of the argument have such heavy support in world culture that they can, without fearing a label of arrogance, dismiss the views of creationists as quackery. Consequently, creationism does not offer effective opposition to the implication of the theory of evolution that there is no need of God. In fact, it provides materialists with a convenient way to equivocate against anyone that challenges Darwinism however plausible their argument might be.

My biggest objection to Darwinism is the way its supporters ignore the problem of the gaps in the POA, and the way in which Darwinism implies humans descending from a sequence of increasingly ape-like ancestors. The fact of evolution is that humans have evolved through innumerable transitional types never shown in the textbooks from a worm-like creature called a Pikaia. The most daring contention shown by evolutionists is to push human evolution back to a lemur, a more primitive metazoan than the chimpanzee. However, the lemur is still light-years distant in complexity from Pikaia.

Nevertheless, the data and its interpretation has provided science with a powerful argument for evolution. I do not deny that in 13.6 billion years enough fortuitous events might have happened that created the POA as we observe it, but having God as the premise of my argument, there is a more plausible explanation of the *gaps* in the POA than Darwinism. That is not to deny the value of restricting science to observation, but to object to the implication that there is no need for God, yes even a "God in the gaps."

The POA seems to bear out that when we see that reality progresses slowly for long periods or even digresses and yet continues to increase in complexity as the result of the

111

actualization of new forms of matter, then there is a purpose of some kind. The theory of evolution is the best explanation for the fossil record during the stasis periods but fails completely as an explanation for the creation events. It is during the creation events both those shown as main events and those not shown but resulted in major new mechanisms such as language, gender, and parenting and in material morphological components such as jaws, legs, and wings during the stasis periods, that God was most likely to have intervened.

4. The Biological Path

The biological path through the sentient stasis period was one of intense evolutionary changes. It began with insect-like creatures living in the seas and other bodies of water and it ended with the creation of the first human mind. Throughout the stasis period, life evolved more complex organisms using and modifying the core processes and adding new morphological changes such as teeth, four legs, lungs, smell, jaws, bones, shells, backbones, cerebral cortex, wings, and amniotic (shelled) eggs while utilizing core processes made possible by the retention of the basic form of DNA. The path of human evolution throughout the sentient stasis period is summarized in figure 9. The chart lists only the main line of clades. For example, before metazoans evolved to the chordates the biological path passed through several such as Eumetazoa to ParaHoxozoa to Bilateria to Nephrozoa to Deuterostomia. It is through these transitional clades that the main human line separated from other lines of metazoa like jellyfish, insects, spiders, centipedes, and shrimp. Cambrian fauna included the basic features that define the major branches of the tree of life from which later life forms emerged. For example, vertebrates are part of the Chordata group found in the Cambrian record. The chordates are related to vertebrates based on the presence of a nerve cord, gill pouches and a support rod called the notochord. The living groups of vertebrates appeared much later. Although Cambrian organisms appeared near the base of major branches of the tree of life, they did not possess all the characteristics of modern animal bodies such as legs. These defining characteristics appeared sporadically over a much longer period. In the 500 -million years of this stasis period, scientists found fossils in the strata piled on top of each other in chronological order that tell a story of new species appearing, radiating, and disappearing. Before the Cambrian fauna crawled out of the seas, they first had to develop the ability to survive in a new environment. This implies that some other genetic deviation enhanced the transition. The main biological path that led to Homo sapiens (modern humans) began during that Cambrian Explosion with the small worm-like creature called Pikaia.

The point of the chart is that the simple Pikaia transitioned through many forms before the emergence of the first humans. The major animal body plans that appeared in the Cambrian Explosion did not include the appearance of modern animal groups such as starfish, crabs, insects, fish, lizards, birds, and mammals. These animal groups all appeared at various times much later in the fossil record than the Pikaia. Figure 9 records the sequence that evolution followed. The point of the chart is that the simple Pikaia transition through many forms before the emergence of the first human. The major body plans that appeared in the Cambrian Explosion did not include the appearance of the modern animal groups such as starfish, crabs, insects, fish, lizards, birds, and mammals. These groups all appeared at various times much later in the fossil record.

Mya	Main Steps	Main Subgroups	Branched Off Species
590	Metazoa		Pikaia, Trilobites
530	Chordata	Olfactores + Cepalochordate	lancelot
520	Olfactores	Craniata + Tunica	sea squirt, sea pork, sea tulips
510	Craniata	Vertrebrata + Agnatha	tunicates, lancelots
505	Vertebrata	Tetrapoda + fish	lampreys, hagfish, sharks, rays
395	Tetrapoda	Amniotes, + Amphibia	frogs, salamanders
340	Amniotes	Synopsids, + Sauropids	reptiles(dinos), birds, snakes, crocs
308	Synopsids	Mammalia	
11 Clades			
220	Mammalia	Theria, Monotremes	platypus, echidna
160	Theria	Eutheria+Metatheria (marsupials)	possums, kangeroos, wombats
125	Eutheria	Boroeutheria, Atlantogenata	anteater, armidillo, elephants
110	Boroeutheria	Euarchontoglires + Laurasiatheria	bats, hyena, daogs, bears, horses
100	Euarchontoglires	Euarchonta, Glires	rodents, rabbits
90	Euarchonta	Primatomorpha+Scandentia	treeshrews or banxrings
85	Primatomorpha	Primates, Demopters	colugas
75	Primates	Haplorhini, Stepsirrhini	lemurs, lorises
63	Haplorrhini	Simiiformes, Tarsiformes	tarsiers, simian monkeys
40	Simiiformes	Catarrhini, Platyrrhini	new world monkeys
30	Catarrhini	Hominoidea, Cercopithecoidea	old world monkeys,
28	Hominoidea	Hominidae, Hylobatidea	gibbons
15–20	Hominidae	Homininae,Ponginae	orangutans
14	Homininae	Hominini, Gorillini	gorillas
7.8	Hominina	Homo, Australopithecine, Pan	chimpanzee, sahelanthropus
3.7	Homo	H.austrolepithicus	
2.8	Homo habilis	H.gautengeensis, H.georgicus	
1.8	Homo erectus	H.neanderthal, H.ergaster	
.7–.3	Homo sapien	H.heidelbergensis, H.denison	
0.16	H. Sapien Idaltu		
0.05	H. Sapien Sapien	Cro-Magnon	

Figure 9. Path of Human Evolution

The forms that appeared in the Cambrian Explosion were more primitive than these later groups, and the theory of evolution explains all the transitions and innovations from the Cambrian fauna. Nevertheless, time passed, and the genetic line arrived at an animal that we call Homo sapien idaltu, a result of the evolution of the homo line that had many of the physical traits as the Homo Sapiens sans the ability to think. In examining the literature, one finds much uncertainty in finding a direct line from the genus homo to the first true human being. Species and sub-species overlap and there are questions such as is the H. Heidelbergenis in the direct line between H. sapien erectus and Homo sapiens?

I believe that the progenitor—the first H. sapien with a mind—an Idaltu, or a member of some closely related family of sapiens, that acquired the language faculty and mated with an Idaltu of the opposite gender to start humanity; it does not matter whether the progenitor was Adam or Eve. Their children then mated with the Idaltu, and humans spread. This scenario resolves the theists' dilemma posed by the question regarding whether the children of the progenitor was the first animal to think, and thinking requires a mind. It is the origin of the mind that separates consciousness from material complexity

D. Psychogenesis—Common View

The long sentient stasis period had progressed for 500-million years during which metazoa diversified and increased slowly as evolution worked its wonders on the animal body plan in which material complexity concentrated in the brain. Then a new species appeared, a species that was not physically different from the species from which it emerged, but it exhibited a wider range of behavior; it was the first Homo Sapien, a human, a thinking animal. We humans emerged from a sequence of animals of slowly increasing consciousness and followed that path throughout the sentient stasis period. Then, about 50-kya, with a dramatic increase, consciousness separated from material complexity to introduce psychogenesis, the fifth creation event. The first human appeared, and the sapient stage began. Evolutionary science now implies that the first modern human (Homo Sapien) emerged from an existing species sometimes referred to as an anatomically modern human (AMS). Hence, the first human (the progenitor) was born of non-human parents. The change from sentient genus *homo* to the *Homo Sapien* in a line of morphologically similar *hominins* manifested as an abrupt leap in consciousness. The materialistic evolutionists contend that the mind did not appear suddenly, but once it emerged it spread rapidly. However, according to the theory of evolution, a new species starts with a single mutated genome in a single being. Because the mind is an emergent property of the brain, nothing in morphology had to change, and there is no single acceptable philosophical or scientific description of how and when the human mind made an appearance, in fact, there is no acceptable common view of the nature of the mind. That does not mean that there are no hypotheses, in fact, there are many attempts at explaining the mind. The common view simply describes the mind as an emergent property of the brain. There is no arguing with the idea that the mind emerges from the brain. However, using an abstraction such as emergence describes the mind does not explain it. This is the materialistic equivalent of the theists' "God did it." Given "emergence" as an abstraction, materialists proceed down a path to nowhere; but to believe otherwise is to lose membership in the club.

There is no need to describe the common view of psychogenesis because science does not recognize that such an event not only did not happen but could not have happened. The common view is that the mind was a product of evolution that appeared when the brain began to think. Well, no, science does not say that the brain thinks. The paradigm is that consciousness, mind, and thought are interrelated, and in some still unsolved way, emerge from the brain. There have been many attempts at solving the problem of these subjective experiences, attempts at finding a widely accepted explanation of the relationship between the subjective experiences and the brain.

A large part of the problem is posed by materialistic science's refusal to consider the possibility of a psychical substance because it is not directly observable or measurable. On the other hand, science is willing to invent dark matter and dark energy because they help to describe certain observations even though they are not observable or measurable. Why then do they not accept the existence of a psychical substance that may answer many of the unanswered questions? The answer is that the adherence to the philosophical principle—logical positivism—does not allow the belief in the existence of God. However, the presence of a psychical substance might explain phenomena in the same way that dark matter and other abstractions use plausibility to explain otherwise inexplicable observations. Because, my premise, that God exists allows the existence of a second immaterial element, the *mind-body problem* is not beyond explanation.

E. Psychogenesis—My View

After a long sentient stasis period during which consciousness tracked material complexity, the mind appeared as the result of psychogenesis and consciousness separated from material complexity and made a quantum jump. I contend that the mind appeared when modern humans began to think. To think, a human needs language and hence thought is related to the origin of language. The first Homo Sapien, the progenitor, could think, which means he/she manipulated words. Words, complex sounds, with specific meanings, had obviously evolved among the progenitors' species. Hence it was the morphological change associated with the use of language that allowed consciousness to separate from the material body and make the quantum jump in consciousness without an apparent increase in material complexity. (see dotted line in figure 3).

The change in complexity was not as great as the change in consciousness. Since sapience emerged in a human having the same morphology, sustenance, and propagation prerequisites for life that the species from which it emerged, with one small exception in the brain, the introduction of the language faculty. I need only to examine the phenomena that caused the jump from sentience to sapience. There have been hypotheses attempting to explain the origin of language, but with little success.

The first human (a Homo Sapiens Idaltu?) was endowed with the language faculty. It obviously could not have been sophisticated sounds with specific meanings. Language requires a brain; the brain is necessary but not sufficient; sufficiency requires a second principle. The additional element must be whatever creates consciousness because neurons alone cannot be the source of consciousness. Neurons are material, consciousness is immaterial. In my view, *consciousness is the manifestation of the nous*. Therefore, at the explicative level that assumes the existence of God, there is no need to ignore the psychical element, nous, which has been present in the metazoan line since the appearance of sentience with somagenesis.

Surely, the progenitor was not born speaking a primitive language; with whom would he/she speak? Instead, evolution must have been expanding and refining the vocabulary of sounds in a tribe of Homo Sapiens. They must have associated specific sounds (phonemes) with certain objects and later evolved the ability to combine phonemes into morphemes (words) to expand their vocabulary. As the complexity of sounds increased so did changes in the human larynx, and other essential physical requirements. Then at this point the homo sapiens could not overcome the language barrier with a single mutation because language means communication and communication means more than one communicator.

What I believed happened among the precursors of the first human was the association of meanings with visual signs—prosody—and/or sign language. As the sounds became more complex, the vocal system evolved into a protolanguage that evolved words and retained much of the prosody. The transition from complex animal sounds to words, sounds with specific meanings, that when manipulated to form thoughts could only have happened suddenly with the addition of a gene mutation in a single proto human (Adam or Eve). For it is not the meaning of individual words but the ability to communicate relationships between them that constitutes thought. The Idaltu dealt with meanings associated with specific sounds; what God created in Adam or Eve is *syntax,* the arrangement of words in a sentence; it was *semantics* that created the language faculty that deals with meanings. The actualization of the mind and the rapid spread of language resulted in the main innovation, namely the concept of abstraction, words associated with meanings along with the ability to manipulate the meanings stored in the immaterial memory. Appearance in one's mind can be spontaneous —when heard or read— or intentionally when in thought. Language not only freed consciousness from a direct dependence on complexity; it also allowed humans to think and led to the creation of *the mind*, a new kind of reality with a new faculty, the immaterial memory. The mind and immaterial memory, the gift to individuals led to a second kind of reality, *rational reality*. This finishes my discussion of the path of actualization (POA) during which my argument for the existence of God is the presence in the POA of events not explained by the theory of evolution but are explained by the existence of God.

F. Sentient Stasis Period—My View

God's involvement during the 540 million years in which sentient life expanded and diversified, also increased in complexity and consciousness. The observation in the fossil record is of a progression of metazoan life towards increasing complexity associated with increasing cephalization. Materialists deal with this observation by contending that consciousness is an emergent property of the neural maps in the brain. The increasing material complexity combined with increasing cephalization adds emphasis to the idea that that the purpose of evolution is the creation of the human mind.

The theory of evolution explains the creatures that appear in the sentient stasis. However, when a mutation or any change in the environment produces a new *phenome*, the actualization of a possibility has taken place and establishes the availability of that *phenome* of further speciation. If the mutation fails, there is no possibility for the incipient phenome. Either way, speciation depends on mutations. Once the first self- replicating cell appeared during abiogenesis, the contingency associated with the holographic effects caused small deviations in certain cellular frames causing diversified cellular life and provided the elements that created multicellularity. This process was vastly slower than the abiogenesis process, hence, the 3-billion-year biological stasis period. When deviations associated with cellular s-frames that appear as germ cells, produce mutations. Hence, God's design of reality included the need for contingencies that eventually produced a mind and its associated property, free will. In addition to the significant increase in material complexity, sometime during somagenesis, consciousness increased as material complexity (cephalized) as shown by the dotted line in the POA chart. Consciousness increased as higher order of metazoa evolved until God created the first human at which point the consciousness separated from material complexity and experienced an enormous and sudden increase.

Note 28: A phenome is a set of observable characteristics and traits that typify an organism such as skin, hair, and eye color, morphology, and behavior.

Chapter 18—Rational Reality

Relative to the immensity of the universe, we human beings are insignificant in space and time, but we are unique because we think, and we wonder. We can imagine galaxies, the mysteries of the stars, and the beauty of life. I can listen to music, taste a fine wine, see the sun set, smell a rose, touch the baby's soft cheek; I can sleep, dream, run, sit, think, wonder, and do magical things stars cannot. I am a human being trying to figure out why I am here, and why God gave me all these magnificent gifts? And why there are persons that do not believe in God because I cannot accept that all this magnificence is the result of a statistical accident. We share with the stars, through the commonality of matter and motion, that inert duo that orchestrates the magnificence of the heavens. We share the same elements, time, space, atoms, quarks, discrete points. However, I am more. I am being, living, knowing, becoming. This wonderful life is a gift from God, and I wonder why materialists use science as a reason to believe that there is no God.

My intent in Part I was to propose an alternative reality, one that shows how God creates and sustains at a deeper explicative level, instead of the reality that we experience, and science can only describe at a descriptive level. I explain how the two modalities of space —discrete and continuous—can be used to describe a dualistic reality with which to create a form of a hylomorphic model of how God might create and sustain reality. I propose a model of dual reality (MDR). that shows how God might operate at the explicative level. The MDR describes: (i) a reality based on discrete instead of continuous space; (ii) algorithms instead of mathematical equations, and (iii) information instead of energy as the impetus for motion. a single holonomic mechanism instead of the disparate models currently used by science. The holonomic mechanism utilizes information from the cosmic algorithm, and divine impetus operating at the explicative level as the source for that which appears at the descriptive level as the dynamic elements, energy, and time. By hypothesizing a mechanism with which God sustains the universe at the explicative level of reality that is independent of the descriptive level of science, I present a counter argument to scientism, the belief that only science can find truth.

In Part 2: I describe the becoming of subjective reality as one of increasing complexity-consciousness. I depict the history of the universe as the path of actualization, a slowly increasing plot of complexity-consciousness that exhibits five discontinuities: cosmogenesis, geogenesis, abiogenesis, somagenesis, and psychogenesis that successively represent the creation of: (i) the universe, (ii) earth, (iii) cellular life, (iv) multicellular life, and (v) the human mind. I apply the MDR to the path of actualization to explain the creation events while accepting the theory of evolution as the description of the stasis periods. In using the same holonomic mechanism to explain both physical and psychical reality, I strengthen my argument against scientism by demonstrating that a God driven mechanism is more comprehensive for explaining reality than is science's disparate, multi-theory approach.

Each human is part of a larger current of humanity that is advancing individually through personal development and collectively through history. Humans are living, knowing, becoming creatures. The first thing in finding the answer to the question, what does it mean to be human, is to address the first mystery, what is life? And even more mysterious is the question, what is the mind? I provided—given the presence of a psychical component definition of life and finished Part 2 with a discussion about consciousness and subjective reality that led to psychogenesis, the actualization of the human mind.

In Part 3 I apply dualism to introduce the human mind, its relationship to consciousness, and its impact on human behavior. I use the presence of an immaterial substance to explain psychical phenomena that materialistic science cannot explain and as a result I add plausibility to my premise that God exists. Then, using my view of consciousness and the mind, I recognize three levels of human behavior based on three levels of motivation: (i) basic instincts that motivate animal behavior; (ii) tangential insight<14>; that motivates rational behavior; and (iii) radial insight that motivates transcendental behavior. The result is the human animal, the rational human, and the spiritual human. However, the most important idea introduced is that of an immaterial memory that allows explanation of sapience, the ability to think.

A. Immaterial Memory

The "God-exists" premise allows a hylomorphic structure of reality in which the two modalities of space lead to the existence of nous, an immaterial substance. Nous is the substance in which an immaterial memory can be created to provide a means for explaining objective, subjective, and transcendental reality. To describe the mental states, faculties, and functions with which the mind operates, a model of immaterial memory is presented.

The key to understanding the mind is to understand the nature of the elements and how they are organized in the immaterial memory. Mental elements are shown in the immaterial memory to provide two functions: sentience and sapience. The sentient is an afferent (input) faculty in which mental experiences called qualia are generated by stimuli external to the mind. The sapient is an efferent (output) faculty in which mental representations such as cognition, thought, and thinking occur in the nous where they can function as internal stimulus of somatic behavior. The sentient faculty is a body-mind function that is the source of consciousness and afferent events and induces our basic instincts that produces human animal behavior. The sapient faculty is a mind-body function that provides tangential insight that induces rational behavior. Whereas the sentient faculty is a passive element, the sapient faculty is an active one.

The main effort in the neuro-science community, the common view, is directed at a body–mind problem centered on consciousness. My view, otherwise, is centered on an immaterial memory that functions in both afferent and efferent ways. This provides plausible explanations not only for consciousness but also for mind, and the mental functions such as qualia, volition, and intentionality that are so puzzling in the common view. I begin the description of immaterial memory with a faculty that operates on both sides of the mind, free will

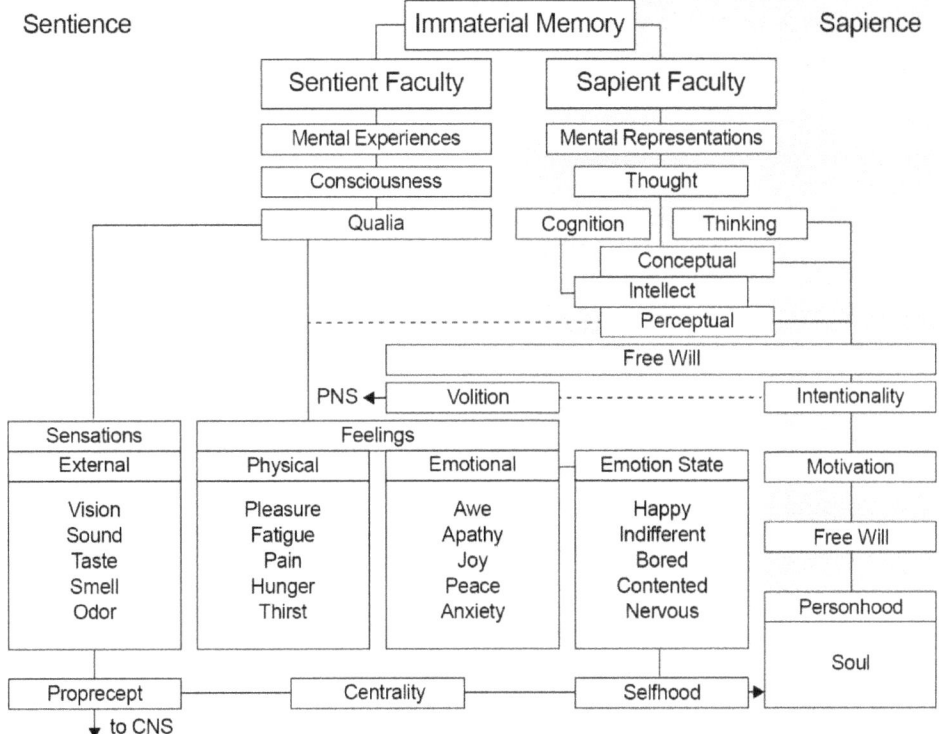

Figure 10. Immaterial Memory

B. Sentient Faculty

The sentient state resides in the immaterial memory where it creates subjective experiences through the peripheral nervous system. Sentience can be thought of as input faculty in which information that flows into the brain activates qualia. In the common view, sentience is an emergent property of the brain. There can be no doubt that the brain is necessary, and in keeping with the use of abstraction for explaining observed mental phenomena, science seems satisfied with the abstraction of "emergence" as the explanation of sentience. What is missing in the common view is an answer to the question, what is emergence? I offer an explanation in my view, but it requires the belief that there is an immaterial memory in each human being; and the immaterial memory requires a hylomorphic dualism in reality; and a hylomorphic reality implies the existence of God. So, before addressing emergence and other mental experiences I will describe my view of the mental elements in the immaterial memory. The sentient faculty is associated with mental experiences. The sapient faculty is associated with mental representations. Mental experiences are simply the sensations and feelings we experience - mostly when we are conscious. Consciousness is the background in which they reside. They form the immaterial memory and are stored in the nous as qualia.

C. Consciousness—Common View

Of all the mysteries in the universe, the faculties and the functions of the mind rise above all the rest, especially that of consciousness. It is perplexing because we all have an intimate experience of consciousness, and experience is invariably more expedient than explanation. The main problem for understanding consciousness is the indecision as to whether it is a faculty, a function, a state, a property, or a substance and whether it is monistic or dualistic. Hence, relying on consciousness as the key to understanding the mind, is the wrong road to travel. A better approach is to unravel the mind by identifying its individual elements and then examine each one's nature and function, then the mystery of consciousness might be solved.

The first problem generated by consciousness is its definition. The consensus I gained from my reading and listening to presentations by philosophers and neuroscientists reduces to the simplest definition found in dictionaries as expressed in two pithy statements: (i) *consciousness is an emergent property of the brain;* and (ii) *consciousness is the state of being conscious.* Statement (i) is irrefutable, because we know that the source of consciousness—because we experience it—is our brain. The problem is dealing with the word "emergence," a classic scientific abstraction that describes but does not explain. Statement (ii) sums up the status of what is known about the nature of consciousness, namely, next to nothing. In general, the more intimate the experience, the more difficult the explanation. The two definitions shown above describe consciousness as a property of the brain or a state of being. As a property of the brain the explanation is shifted to the abstraction "emergence." In fact, the word "consciousness" itself is the most convenient and misused of all the abstractions. What is it, into what does it emerge? As a state, how does consciousness relate to the three main states of being, namely alive, awake, and aware? It is most often conflated with awareness, a short cut of stretching definitions; awareness and consciousness are not the same. But property and state are not the only categories in which we find consciousness; it is also thought to be a faculty or a function.

A faculty view addresses what consciousness is; the functional approach addresses what consciousness does. As a faculty, the two main views of consciousness are monism and dualism. The monism/dualism faculty problem is discussed in the previous section.

1. As a function, Ned Block (1942–) distinguishes a phenomenal (P-consciousness) and an access (A-consciousness). P-consciousness involves the experience of qualia (sensations and feelings). A-consciousness involves information in the mind regarding speech, reason, and control of behavior. It functions as perception, introspection, and recognition of information.

2. A second functional view, proposed by the philosopher David Chalmers, is that the mind-body problem consisted of the hard problem and an easy problem. The hard problem is: how does the body induce subjective experiences such as sensations, feelings, and emotions (qualia) in the mind? The easy problem is mapping the precise parts in the brain that are associated with each physical and mental representation.

The literature refers to both hard and easy problems of consciousness as the mind-body problem. This view treats consciousness as a basic "substance" on which all other subjective experiences are related. In addition to the faculty and functional approach, consciousness is examined by its nature as a: fundamental substance such as: integrated information; a quantum state, sentience; awareness; sapience; knowledge; and the unique views that shows consciousness can be almost anything you want to make of it, for example, even philosophers with great reputations offer these:

1. James, William (1842–1910)
    "Consciousness is not an entity, not a thing, but a flux and system of relations; it is a point at which the sequence and relationship of thoughts coincide illuminatingly with the sequence of events and the relationship of things." —Quoted in: *The Story of History*, page 383, Will Durant, 1953

2. Schopenhauer, Arthur (1788–1868)
    "Consciousness is the mere surface of our minds, of which as of the earth, we do not know the inside but only the crust." —*The World as Will and Ideas*, London, 1883

D. Consciousness—My View

The brain is necessary as the material part (language) but is not sufficient, and an immaterial memory is also necessary but not sufficient; the mind requires the presence of both. Since I define the mind more directly as an interconnection of material language faculty of the brain with an immaterial nous, we can define consciousness as a manifestation of the nous, the immaterial substance. Mind and consciousness are separate entities.

When the brain dies, the mind, consciousness, feelings, emotions, thought, and life disappear, so the brain is necessary to have subjective experiences. Based on that observation, materialistic scientists assume that the brain is the only active element. In the common view, the general approach is to explain consciousness and the other explanations of subjective experiences will follow. Hence, the common approach to the mind-body problem concentrates on consciousness, and the prevailing paradigm contends that all subjective experiences are emergent properties of the brain. That is certainly true, the mind emerges from the brain, but "emergence" does not explain what the mind is.

I describe my view of the mind earlier and have come away from that, equipped with a set of tools, especially nous and immaterial memory, which allow to find more plausible solutions to the so-called mind-body problem. Remember, in my view, the mind consists of the language faculty in the brain combined with the psychical substance, nous, in which the immaterial memory is formed.

E. Consciousness as a State

Consider the possible states of conscious metazoan life: *alive, awake,* and *aware*. Suppose you are alive, not awake, and not aware, you could be asleep comatose, unconscious, but still *sentient*. Suppose you are alive, awake, and not aware, you could be drunk, in a stupor, traumatized, but still *conscious*. Suppose you are alive, awake, and aware, you would be *sapient*, in other words, human. If the brain functions, an animal is alive and sentient. When stimuli, in addition to activating the brain, also activate the immaterial memory, the animal awakes, and we say that it is conscious. Sentience is associated with being alive; consciousness is associated with being alive and awake, and sapience is being alive, awake, and aware. Hence, sentience means to be alive; conscious means to be awake; and sapience means to be aware (to think). *Alive* means the material memory is active; *Awake* means the immaterial memory is active; *Aware* means the mind

121

is active. So, we can say that to be conscious is the *state* of being awake, and consciousness is the experience of being awake. I am playing a very subtle word game, but what I am trying to establish is a difference: between an act and a state; for example, between conscious and consciousness, or between awakening and awakened. This is not unlike the difference between thinking and thought, in which thinking is an act and thought is the result. I have arrived at the idea that to "be conscious" simply means to be awake, and that kicks the ball down the street to the real problem, what does 'being awake' mean, and what is consciousness?

1. Awake

To be awake means to be conscious; to be in full realization of one's senses. Sight is the one sense that is completely inactive when asleep—the other senses are still active but diminished. What we experience when we awaken, and we become fully conscious, is a panorama of all the senses, including sight and sound that are in remission when asleep. It is this panorama of sensations that is what we mean by consciousness. In the common view this panorama is referred to as qualia, in my view, consciousness is the panorama of sensations. And, since the sensations exist in the immaterial memory, consciousness is the manifestation of the nous. Hence consciousness can be treated not only as the state but also as a property.

F. Consciousness as a Property

Consciousness is an emergent property of the brain. Emergent simply means it comes from but does not reside in the brain. But what kind of property? We might say: *consciousness is to X, as wetness is to water.* X is, of course, *nous,* a psychical substance proposed 25-centuries ago by the Greek philosopher Anaxagorus. Consciousness emerges from the material brain when the material brain engages in the immaterial nous of the mind. It is the nous, not consciousness, which is being manifested; consciousness is the experience of the qualia imbedded in the nous. Because nous is a psychical substance, the common view disallows it because it leads to the existence of God and that is not part of the modern scientific paradigm. However, the "God-exists" premise allows the existence of nous. The existence of nous allows the existence of an immaterial memory, and the immaterial memory allows us to explain the mind, subjective, rational, and transcendental reality. Since immaterial memory is associated with the brain, it is the most direct way to explain "emergence"— science's view that consciousness is an emergent property of the brain. The brain is necessary because the mind depends on syntax, the words, and their arrangement, stored as a language faculty as part of the material memory. The meanings generated by language (semantics) reside in the immaterial memory. Hence the mind emerges from the brain but only in association with immaterial memory. The existence of immaterial memory answers the question regarding emergence, namely, where do the properties emerge to? The emergent properties "emerge to" and are stored in the immaterial memory. Because of nous, a substance that forms the immaterial part of the mind, the explanation of conscious is straight forward. Mind and consciousness are separate entities. The mind is a faculty; consciousness is a property of the mind and a state of the brain. Consciousness is associated with the sentience system in the immaterial memory.

In the brain, when a neural map associated with a specific quale is activated, it induces the experience in one's mind. The result is a motion picture inside our head, we call it consciousness. The bottom line is that consciousness is simply the experience of qualia. The mystery of consciousness is simply conflation with the word qualia and the suppression of the concept nous, the true fundamental element of our mental world. An explanation of consciousness may solve the body-mind problem, but there are two main functions of the immaterial memory, consciousness is the input function, sapience is an output function. There is also a need to explain sapience, the output side of the of the immaterial memory. Sentience is called the hard problem of consciousness because of the inexplicable relationship between the neural map (the body) and qualia (the mind). I see sentience is an afferent body-mind problem and sapience is an efferent mind-body problem.

G. Qualia

Qualia are innate vibrant experiences stored in the nous of immaterial memory. They are passive in that they require an input from the neurons in the brain to "bring them to mind." This body-mind faculty involves events in which sense receptors send impulses through the afferent nerves in the central nervous system to the brain where neurons are activated. A conversion from digital to analog takes place. The experience of qualia is strictly subjective. We can never know that our qualia are the same as that of another person's despite a common agreement about what we experience.

Qualia are a direct gift from God and are innate vibrant properties of the nous experienced through the senses. I use the word vibrant to describe the nature of nous, meaning that nous, a continuous spacelike substance, is not empty but exhibits the capability of producing our subjective experiences just as discrete space exhibits the capability of producing a magnetic field. We can think that just as the phenomenon of magnetism exists in a magnetic field formed in discrete space, the sensation of color exists in the immaterial memory in continuous space. The action of matter induces both qualia and magnetism; neurons induce qualia; electrons induce magnetism. Science deals with qualia simply by making them "emerge" from the neurons in the brain. Qualia, such as light, exists in all metazoa that exhibit consciousness. In any experience of light, there certainly is an uncountable number of photons, but photons only become light when seen by eyes and interact with nous. Although the experience of qualia is found throughout metazoa, there are, among people variations, for example, color blindness, enhanced taste and hearing, pain tolerance are among the variations. The variation among species can be great, dogs do not experience red and green for example, while mantis shrimp have trinocular vision, therefore can see more colors than people. Qualia also becomes part of a percept through volition (explained later).

Note 29: It is interesting to note, the universe existed in complete darkness for 13+ billion years before light appeared with the first organism with eyes.

1. Sensations

Sensations are qualia that originate externally as environmental stimuli such as light, sound, flavor, odor, heat, or pressure detected by sense receptors and processed in the material memory and terminate in an afferent event that creates a subjective experience in the immaterial memory. There are two types: (i) sight and sound can be formable; (ii) the rest of the senses and all the feelings are amorphous. The sequence: source-sense-category-qualia represents a sensation channel. For example, an apple reflects light in the form of photons for processing in the material memory where they generate events, and we perceive them in our mind as a spherical, smooth, red object. Each quale follows a similar sequence: (i) from source; (ii) to the *neural map* associated with the sense; (iii) to the afferent synapses that "bring to mind" the redness of an apple; hear the music of the violin;

smell the aroma of a rose; taste the sweetness of honey; or touch the smoothness of glass. We perceive our senses and store them as percepts. Sight and sound perceive formed percepts meaning that their source includes forms generated in the material memory. With sight, for example, *retinal ganglia* in our eyes create forms that combine with color to form a perception that is stored in immaterial memory. How color combines with form is called the *binding problem*.

2. Feelings

Feelings are qualia that originate in internal sense receptors in response to physical (homeostatic) or emotional (endocrinal) demands. Unlike differentiated sensations, feelings are amorphous. Common physical feelings (figure 10) are in the larger box. Obviously, there are other experiences that belong in those hanging boxes, but I leave it to the reader. What I include in each box should suffice to cover a wide range of experiences.

Physical feelings are those subjective experiences that originate internally at sensors associated with *homeostasis*, the self-regulating mechanisms that are part of the autonomic nervous system that provides the optimal internal environment for the animal's body to function. In general, the cause of physical feelings is known and remedied. If we feel pain we attempt to eliminate or avoid the cause; if we feel hungry, we eat, if we are thirsty, we drink. There are, of course, positive physical feelings, but they are not as easily distinguishable as the negative ones. Correction of a negative feeling induces a feeling of well-being that is the same relief no matter which of the negative feelings we negate. Going beyond satisfaction leads to the creation of levels of pleasure that go from comfort to satiation.

Emotional feelings are those subjective experiences that are short-lived, focused, negative or positive, the source of which is external or internal and have specific causes. For example, the causes of negative emotions such as anger, jealousy, contempt, and disappointment exist external to one's corpus. The cause of negative emotions such as frustration, shame, embarrassment, and regret exists internally within one's mind. A positive emotion such as pride, pleasure, satisfaction can arise externally. A positive emotion such as awe, peace, and joy can arise internally. We experience most emotions in a variety of modalities over a span of intensity from positive to negative. For example, personal relationships could range from: love to admiration to toleration to annoyance to disgust to hate, with several half steps and divergent paths in between.

Emotional states are the culmination of one's emotions. Whereas an emotional feeling is short-lived and intense, an emotional state persists longer and is milder. When experienced for a brief period, we often internalize an emotional feeling; but when experienced intensely can become a *mood:* and when it becomes unavoidable it becomes *temperament* and becomes part of one's personality. Moods can be positive or negative. Moods arise from emotions, they also arise from other factors such as lack of sleep, lack of nutrition, disappointment, failure, aggravation on the negative side, and success, good news, and friendship on the positive side. Humans react to the difficulties of life in diverse ways according to their basic nature.

H. Selfhood Path (v)

Selfhood is one part of a human's personality; the other is *personhood*. Selfhood, a mental faculty, is formed as the result of two influences: basic instincts and one's innate emotional state. The path contains two faculties: (i) the sense of proprioception; and (ii) *centrality*, the focus on self.

1. Proprioception

Often considered as a sixth sense, proprioception is a special sense shown at the bottom of the immaterial memory chart (figure 10). It is experienced in two modalities: (i) internally, it is the sense of being aware of the position of each part of the body relative to the other parts, it is the sense that allows you to touch your nose with a finger; and (ii) externally, it is the sense that allows you to know where the body is in relation to one's surroundings—hence one can imagine, in general, the sequence one must follow to get from where one is to where one wants to be.

> Note 30: The common view sometimes adds *kinesthesia*, the study of bodily movements and the vestibular, the sense of balance as part of proprioception.

Internal proprioception is the awareness of position and applied force of the parts of one's body derived from sensory inputs from muscles, tendons, and joints. Proprioception arises in the nervous system by sensing muscle action and balance provided within the inner ear and stored in material memory. Internal proprioception is important because it allows a person not only to locate the parts of the body but also is essential in guiding their movements. For example, fingers will automatically apply the right amount of pressure to pick up objects of various weights. Sensory feedback from the sense receptors in the fingers adjust the muscular action to apply the correct pressure to the object. Neurologists have thoroughly described this form of proprioception. However, proprioception will also arrange the fingers differently before contact with the objects as different as an apple or a pencil where before the fingers touch the object and the muscles react to the sensory feedback, the fingers know which kind of object is involved. In other words, science cannot explain proprioception solely as pressure exerted on sense receptors, something else is involved.

External proprioception is important because at any moment an animal is an integral part of the configuration of s-points that comprise the entire universe, the so-called cosmic s-frame. Since proprioception is a source of awareness, proprioception can very well be the most important sense because it allows a person to "be aware" not only where every part of their body is in relation to all other parts, but also to be aware of the milieu—where the body is in relation to the rest of the universe.

Because animals do not store as many concepts (words) because they do not have language, they have higher perceptual intellect than humans and their sense of location within the cosmic s-frame appears to be higher, if not exceptional. (This may be the answer to the mysteries of animal navigation). However, humans have a discriminating sense of proprioception because we know where we are most of the time. Where one is born, the country, the nationality, the family, are important factors in the development of one's soul. All these factors contribute strongly to our identity that often predominate basic instincts, hence our sense of self, our selfhood. We all understand the impact our procreation, i.e., our birth rite, has on how we think and how we behave, but there is another factor that strongly impacts our basic nature that is independent of our procreation; I call it *centrality*.

2. Centrality

The center of a spherical volume is a point that is equal distance from its surface. If, as science contends, the universe is infinite, then every point within the universe is equidistant from the infinitude and there is an infinity of centers. The modern scientific paradigm views this as equivalent to stating that the universe has no center. Or, because the universe is so big that only the local expansion is observable, and every point is moving away from every other point, i.e., that every point is at a personal center of the universe. I contend that the physical universe is finite, so science's view that every rational point is at the center does not apply; but because the cosmos is infinite every real point is at the center of the cosmos. This means that no human bodies are at the center of the universe, but every human soul is at the center of the cosmos. This is what I mean by centrality. Centrality generates selfhood,

Innate emotional state is the other clue to the presence of selfhood in the newborn and is the observed manifestation of emotional states, i.e., moods and temperament. Most psychologists do not believe in the innateness of temperament. But in my experience with 8 children, 22 grandchildren, and 23 great-grandchildren I can think of no other way to explain their diversity of early temperament than through inheritance. There are two factors contributing to the inheritance of selfhood: (i) the DNA physical factor; and (ii) *traducianism*—a psychical factor.

DNA, the physical element of selfhood, emerges when the zygote inherits the physical pattern through DNA from both parents, as a result, babies are born with an inestimable diversity of phenotypes (morphic s-frames). Physical characteristics are a significant element in one's selfhood. Humans are born with a variety of physical attributes that can impact their selfhood. Gender has its impact, but so do traits such as race, stature, and intellect. These are traits derived from one's DNA and in some ways influence the basic instincts, especially the need for identity. In the common view, DNA and gene expression not only explain the physical but also the psychical characteristics. It is easy to accept that the physical characteristics in each person were the result of the blending of the parental DNA, but what about temperament? In the common view, temperament is the result not so much of nature but mostly of nurture. Although there are those that recognize that predisposition to temperament is innate, the common view is that it is related to physical characteristics. This raises the question as to how does DNA produce a diversity of temperament? I contend that it does not, but the psychical substance does.

At conception, the bios from both male and female gametes combine to form the nous that corpuspirates the zygote. In Christian theology this is known as the doctrine of traducianism. It means that the complete person, body, and soul, are transduced by natural generation. The bios contained in the gametes imparts an innate selfhood to the fetus and babies are born not only with a predisposition, but an incipient temperament. Hence, some babies are born happy, some are not; some are born peaceful, some are not; some are wonderers, some are not; some are obstinate, some are not, and the list is long. Since selfhood is a manifestation of the basic instincts, and since the instincts are permanent faculties, selfhood is a permanent faculty of human life.

I. Selfhood—{s}

Selfhood is a mental faculty that deals primarily with our focused individual needs, the basic instincts. Selfhood is the foundation of human nature, the nature with which we are born, the nature no one else is privy to. In humans, the need for identity predominates. The need for identity derives from centrality that derives from the sense of proprioception. However, in infancy, the need for security exceeds identity. A hungry baby lets the parent know quite forcibly with ear-spitting screams to express its need for security (sustenance). A need for stimulation initially subdued soon comes alive when the infant responds to pleasant sounds and comforts with smiles and cooing. Both security and stimulation needs are self-directed, so the need for identity, even when not expressed directly, happens in infancy indirectly through other needs. Selfhood derives from the action of the sentient faculty, the left side of the immaterial memory.

This ends the discussion of the sentient faculty, the body-mind, afferent mechanism, and we turn to a discussion of the sapient faculty, the mind-body efferent mechanism. The sapient faculty is ignored or trivialized, in the common view, as the "easy problem of consciousness." I now turn to the sapient side and the second of two main immaterial faculties, the sapient faculty.

J. The Sapient Faculty—{I}

The sapient faculty is an idea ignored in the common view. The sapient faculty: (i) thinks; (ii) gives rise to sapience; and (iii) makes decisions that activate behavior. The sapient faculty performs a dual role: (i) as the site of all the mental representations: cognition, thinking, thought (ii) as the mysterious voice in one's mind that makes decisions and induces mental and physical behavior (free will), and in fact, controls one's destiny. I symbolize the "sapient faculty" by the symbol, {I}, and refer to it as the *Commander I* or simply, I or ME.

Because the common view disallows the existence of a psychical substance like nous, it does not recognize that it is the sapient faculty that is the hard problem of the mind because of *causal interaction*, the problem of explaining how the physical and psychical impact one another. Emergence is not an answer for the control of the physical by the psychical mind; that would require the material to "emerge" from the immaterial. Although immaterial memory is essential, describing the nature of the mental elements and their interactions still presents an overwhelming challenge. In addition, the lack of specificity in words adds to the difficulty. For example, thought is a mental faculty; thinking is a function; what thinking produces are thoughts and thought also is the past participle of thinking. Except for the past participle, there are differences between thought and thinking. In addition, to understand the mind and its functions, we must deal with words such as cognition, consciousness, awareness, intentionality, and intellect.

K. Mental Representation

Mental representations are mental images that represent external reality and are utilized by three mental functions: cognition, thought, and thinking. They can be perceptual or conceptual. One of the distinctions resolved with hylomorphic dualism is the distinction between conceptual and perceptual, here is a personal example for what I mean.

In a book about Artic exploration *The Terror*, by Dan Simmons, I read the following sentence: *"when the temperature sometimes rose above freezing, the skies were filled migrating birds. Franklin himself could identify the petrels from the teals, elder ducks from the little auks, and sprightly little puffins from all the others."* My mind imagined the scene as I read this example of perceptual thought. Then the date of record, September 3, 1846 came to mind, and I wondered how long ago that happened, so in my mind I subtracted 46 from 100 and got 54, then added 100 for 20th century to get 154. Then added the 22 years from the present century and came to an answer of about 176 years ago. This is an example of conceptual thinking (manipulation of words). This simple example of how the mind works effortlessly jumping from one mode to another is a reason it is so difficult to explain how the mind works. It is also an example of showing how hylomorphic dualism—in illustrating the source of mental modes—may offer the path to finding answers to the mental ambiguities.

I contend that finding the mystery of consciousness and the other mental elements is impossible without dualism. A material memory formed from a complex network of billions of interconnected neurons—called the nervous system—is a system thoroughly studied and described by neurobiologists. However, if as the materialists believe, the mind is associated only with the material brain, then they will have to find a way to explain how neurons think! That *is* a problem that is *not* under consideration because materialists have other problems to think about: (i) they still have not solved the problem of consciousness—the problem of explaining how neurons generate qualia; and (ii) have and have hardly addressed the sapient faculty. Of the mind's faculties the somatic and the sentient were discussed above. Here I deal with the sapient faculty and its three elements: cognition, thought, and thinking.

L. Thought

Although thought is not the sole property of the mind, the awakened mind is always in thought. The mind makes both minor decisions such as, *"should I read a book or watch TV,"* and major decisions such as, *"should I join the army?* that require analysis of the situation that is *thinking*, not *thought*. The mind rationalizes when someone or something criticizes or questions your identity, and you go into a defensive mode. The mind can engage in a creative mode through wonder and learning. However, a great deal of time, most minds are in reverie, the free association of ideas that is spontaneously changing subjects. Decisions, analysis, rationalization, wonder, reasoning, and reverie are a few of the activities made possible with thought. We understand the various modes of thought because we experience them directly, but there is no general agreement as to how to explain or define thought, and in my view, there never will be if it remains solely a function of the brain. The best observation we can make about thought is that we think with words or as I will argue later, we think with word images.

Can anyone imagine neurons having thoughts? There is no doubt that neurons are necessary, but they cannot be sufficient; neurons or any combination of neurons cannot have thoughts! The best that materialists can do to describe thought is to contend that it is an emergent property of the brain. What does emergence mean? I can find no explicative answer; science merely describes emergence as something observed. I have used the word "thought" about 50 times previously in this synopsis without defining it, trusting the reader to grasp its use in context. But that does not explain what thought is. Earlier I presented a description when I wrote: "thought is a mental faculty; thinking is a function; what thinking produces are thoughts and thought also is the past participle of thinking."

1. Thinking

Sapience means to think, and "to think' means "*to manipulate word images with which to derive meanings or to reference percepts.* I contend that only with the presence of an immaterial memory can we explain thinking. Thinking is the mental act of consciously manipulating word images (not words) associated with referents (meanings or percepts). Words are created by neurons in the brain, the material memory; word images and referenced meanings are manipulated in the immaterial memory.

Although thinking consumes an unusual amount of our (conscious) hours, when we speak, hear, or read, we are not usually thinking. Speaking, listening, and reading involves both material and immaterial memories while thinking involves only immaterial memory. Thinking is the spontaneous manipulation of word images without a purpose such as reminiscence, imagining, reverie, daydreaming, meditating, study, scrutiny, inference, introspection, and perception. To simplify somewhat I reduce thinking to three general activities: cognition, recognition, recall.

> Note 31: Word images are mental representations of written or spoken words that signify referent meanings or percepts. They exist in immaterial memory as mental images.

Since thinking utilizes words which are discrete, thinking is a conceptual function. The manipulation of word images tends to blend in with meanings much like the blending of phonemes to make spoken words, in effect, we are manipulating the meanings of words as percepts. As we read the words in the previous sentence, we are deriving a conceptual conclusion from the meanings associated with the word images, namely that all thinking takes place in the immaterial memory. Alternatively, a word image can refer to a percept. For example, "Look at the pretty flower!" is a sentence in which word images reference percepts"! With word images, we are thinking of them as our eyes scans the sentence.

Most thoughts are intentional; they are not spontaneous. Thinking manipulates word images as cognition extracts their meanings and merges with motivation, then {I} utilizes volition and decides to employ or not employ (free will) that which initiates or retards action.

Thinking is the manipulation of thoughts towards a purpose such as contemplation, solving, planning, wonder, and any other thinking towards a goal or a purpose. Given the presence of the immaterial memory, we can explain thinking more precisely than if restricted to the neural map of the material memory. When the creation of a language faculty combined with nous in the immaterial memory, the mind appeared with the capability to think, the intentional manipulation of word images. The elements of the immaterial memory—described earlier—form a sequence of subjective experiences that create thoughts. There are three elements in a thought: a signifier, the word image and the referent. The signifier is word that activates a neural map; the **word image** is the unspoken, unwritten version of the signifier that is stored in the immaterial memory; and the **referent** consists of a percept and all the information associated with the signifier. Percepts—as I will explain below—can exist without a signifier, especially for use in the mental activity we call cognition.

> Note 32: A signifier is a mental representation stored in the material memory and is connected to the referent that is stored in the immaterial memory by a signifier image.

2. Cognition (iv)

Cognition is a function that creates perceptual intellect through observation. As the morphic s-frame moves through space and time, absorbing sensations, cognition analyzes the perceptual input, decides its value, and influences the mind. We store few in one viewing. Most percepts stored in nous are either perceived repeatedly or attached to a *signifier*, usually a word stored in the material memory. When activated, either through the ears as a spoken word or read with the eyes, a signifier connects with a *signifier image* that

Note 33: A *signifier image* is a mental image of the signifier that is stored in the immaterial memory as part of the percept or concept and communicates with the signifier stored in the material memory in both afferent direction for recognizance efferent events. A word becomes a *word image* as an element of thought.

is stored in the immaterial memory where it is available as either afferent (recognition) and efferent (recall) directions. The neural map of the signifier stored in the material memory is linked with a signifier faculty image stored in the immaterial memory. The linkage occurs at synaptic terminals and is the faculty that allows the recognition and recall functions. Cognition creates perceptual intellect by selectively acquiring percepts from the external world and storing them as *referent*s consolidating them with a signifier such as a word and storing the perception and the image of the signifier in the immaterial memory. Volition provides the power of storing perceptions in immaterial memory. For example, when a face is associated with a name to form a "conceptual percept." A concept such as a name is stored in the material memory and its image is stored in and connected to the percept in the immaterial memory. The percept is a selected perception plus a signifier image. There are far more percepts stored in immaterial memory than are concepts stored in the material one. For example, I recall far more faces than their associated names, and more melodies than lyrics of songs. Immaterial memory is much larger than material memory because concepts are associated with rational numbers that are discrete; the percepts are associated with the real numbers that are continuous.

Note 34: Rational numbers, and hence concepts, are associated with aleph-null, the first of the infinite numbers; real numbers and hence percepts are associated with aleph 1, the second infinite number. Since aleph-1 is infinitely larger than aleph-null, we might say that there is room in the mind for storing infinitely more percepts than concepts)

Acquisition, unlike qualia that are innate, we acquire intellect: (i) conceptual through learning, and (ii) perpetual through observation. The immaterial memory chart (figure 10) shows two paths from the intellect to its effect on a person's motivation. One path based on concepts that passes through thinking and intentionality; and a second path, based on percepts that passes through cognition.

Consolidation: We consolidate the acquired information by determining which information is worth keeping and storing as knowledge. Knowledge is retrieved in several ways.

Retrieval: An efferent mental function that utilizes the intellect that bring intellectual elements to mind. It appears in many modalities: recognize, recall, remember, recollect, remind, reminisce, evoke, notice, perceive, conceive, imagine, and many others like I see, I get it, I agree, etc., etc., *Signifiers* are individual neurons or a neural map that, for a brief duration unless silently repeated, imprint signifier images in immaterial memory. The person's intellect determines the efficacy of this learning process. The interaction of the signifier with the signifier image impacts the faculty of memory and its two components: recall and recognizance. This interaction appears as intellect. Although both concepts and percepts have similar construction regarding signifier and referent, they differ intellectually. There are two types of intellect, conceptual and perceptual.

Recognition is an afferent function of our mind that is virtually always active as we pass through the day, consolidating innumerable scenes (percepts) that enter our PNS, to become an element in our perceptual intellect where they persist for a time consistent with

130

our perceptual intelligence. The persistence of each percept depends on: (i) the number of times a particular percept is perceived; (ii) whether the percept is connected to a concept; and (iii) the mind's perceptual intelligence. For example, facial percepts (faces) are easier to consolidate than the clothes the person was wearing when first met. My wife has excellent perceptual intelligence and can usually remember a face and the attire of a person at a single meeting. I experience low perceptual intelligence; I am by nature, conceptual. Part of recognition is the experience of the mind in which percepts (faces in a crowd, for example) match a percept stored in the immaterial memory and we say we *recognize* a specific person.

Recall occurs when the sapient faculty brings to mind the knowledge associated with a recognized signifier image. For example, when recognizing a face in the crowd invariably a name is recalled along with any other pertinent knowledge stored with that name. There are other ways to bring information to mind. When the mind is thinking we sometimes remember, recall, reminiscence, recollect, evoke, believe, understand, see, know, or observe.

3. Intellect

A faculty in immaterial memory that stores conceptual and perceptual knowledge. The intellect faculty consists of three functions: *acquisition, consolidation, and retrieval* of knowledge. The reason we often confuse intellect and intelligence is that the common usage of the word intelligence is usually associated with concepts. IQ tests measure conceptual intelligence. Concepts are the tools of scientists, mathematicians, and philosophers, professions requiring persons of high intelligence, and hence, high conceptual intellect. There are other fields that demand an equally high perceptual intellect.

Sense perception ranges from low or none to acute in taste, hearing, smell, sight, and touch, hence the poet, artists, thespians, athletes, and all engaged with intuitive beauty or any other characteristic in which we measure *talent*, exhibit a different kind of intellect, a perceptual intellect, which is no less important than the intellect of scientists, mathematicians, and philosophers.

Intellect is the knowledge stored in the mind; it deals with two types of information: knowledge and opinion. Although there is no absolute meaning for either type, in general, information certified as true becomes knowledge. Opinion is information that could be true but not certified as such. Knowledge constructs the intellect, and intellect gives credence to opinion. In other words, knowledge enhances opinions. This is especially true when one's opinion is based on the same branch of knowledge as one's intellect. Then the opinion becomes expertise. Opinions only count if your intellect is credible, as in "you know what you're talking about." And this is the crux of what I contend because beliefs that I am defending when stripped bare are merely opinions. However, the factor that gives opinion more credence is plausibility, the standard by which to judge the argument between theist and non-believers. And plausibility is what I am striving for, not probability. In replacing probability with plausibility, I diminish opinion by embracing plausibility.

Perceptual intellect is the total amount of information acquired for each perceptual signifier. Percepts are the basic objects of recognition, recall, concentration, performance, and other forms of observation, and of *knowing*. Percepts form the episodic memory. Regarding perceptual intellect, the primary sense is sight. During our waking hours, the eyes absorb innumerable scenes every day, every hour, and every minute. Only those that have value or affect a sensation are stored as percepts. Percepts are amorphous mental representations of reality such as faces, scenes, attire, places, melodies, directions, odors, and other sensations. Percepts are acquired spontaneously through observation or purposely through study or training. They are consolidated with signifiers when necessary and stored in the immaterial memory as perceptual intellect. The standard definition of a percept is: "*all that is sensed of objects.*" My definition is broader: "*all that is sensed of objects in real time and/or all that is known of a signifier.*" Percepts are composed of multiple sensory sources, for example, we recognize others visually by their face, but also by the sound of their voice. Percepts include words and numbers, both by sight and sound, shapes, sensory images, qualia, facts, thoughts, tastes, odors, and often emotions related to a signifier. Unlike the symbolic knowledge of concepts that inheres in the material memory and serve as signifiers, percepts reside in the immaterial memory as referents.

Perceptual intelligence is the efficacy of acquiring perceptual intellect. often referred to as *talent;* it is mostly innate, but to sharpen perceptual intellect we can supplement talent with training (repetition of specific behavior) to become skill. Like all other human traits, there is a broad variation of specialized perceptual intelligence. For example, remembering scenes and personal information is a static form of perceptual intelligence. Other perceptual intelligence can involve time. The time-factor allows one to remember a sequential scene, usually visual or aural. (i) visually like the motion of objects like the arc of a basketball, or (ii) aurally like the melody of a musical piece. Proprioception allows one to remember spatial-temporal changes in sensory scenes that allow one, for example, to dance.

Conceptual Intellect is the conceptual knowledge stored in the immaterial memory as discrete mental representations such as forms, words, numbers, musical notation, and any other discrete mental object that are acquired through learning and are stored in material memory as conceptual knowledge.

Conceptual Intelligence is a mental function that determines the capability of acquiring and using conceptual information, such as words, numbers, names, notes, facts, ideas, concepts, lyrics, poems. Intelligence is often confused with intellect. I prefer to differentiate between the two terms: intellect is a faculty; intelligence is a function. A person can have a high intellect and still act unintelligently; conversely, people with low intellect can act intelligently. From this, we can assume that intelligence is not a strong motivator, its impact is on a person's ability to learn to reason, and to expound knowledge. Learning is the mind's function for building conceptual intellect. On the other hand, intelligence is—as I will discuss in Chapter 20—an important part of motivating rational behavior. At my current age, 90, I am quite familiar with the problem of the efficacy of the mind's capability of transferring what is experienced in short-term into long-term memory, the capability called intelligence. In this regard, I personally have gone, over the years, from an A student to a D student (unless you grade on an age curve).

4. Free Will

Free will is simply the human capability of making decisions; it is realized in: (i) sentience through volition and (ii) sapience through intentionality. On the sentient side there is often a choice to experience certain qualia such as the feeling of pleasure and the sensation of taste. On the sapient side free will always exists because decisions are self-imposed. However, the existence of free will is not absolute. To be able to make a choice freely, the choice must be completely free of legal, physical, psychological, or other restrictions. It is the reason we have rules, laws, and other societal compulsions. Legal restrictions are imposed externally, physical, and psychological are imposed personally. Unless the will is able it is not free. When confronted with situations requiring a decision, the will is still free and can and sometime does initiate action despite motivation. If our behavior always aligned with motivation, the mind would not be free, hence freewill is shown with intention, i.e., behavior that is the result of a thought. {I} decides the form of behavior, consequently free will is the manifestation of choice and we humans can choose between good or evil, right or wrong, rational or irrational behavior, or to act or not to act.

5. Volition

Dictionary definitions for volition read something like: (i) "the act of willing, choosing, resolving; or (ii) the exercise or power of the will." While the definition for will reads: "the act or power of volition." The reason for the conflation of will and volition derives from a materialistic view that denies the existence of the immaterial. My view is based on a dualistic reality that allows a separation of volition and will. Volition is a sentient function connected with the peripheral nervous system and to intentionality, a sapient function. Volition consists of two functions:

a.   When activating the PNS, it induces behavior, for example, (i) when a person perceives a red light while driving and reacts by stopping, or (ii) when feeling hungry and reacts by eating.

b.   When deciding and storing for example, you recognize a friend while walking and immediately recall his identity from your intellect. When the friend introduces his wife, you consolidate her name as a concept. Hence, volition can be defined as the faculty that manages the perceptual intellect.

6. Intentionality

Wikipedia defines *intentionality* as: *the power of minds to be about something: to represent or to stand for things, properties, and states of affairs. Intentionality is primarily ascribed to mental states, like perceptions, beliefs or desires, which is why it has been regarded as the characteristic mark of the mental by many philosophers. A central issue for theories of intentionality has been the problem of intentional inexistence to determine the ontological status of the entities which are the objects of intentional states.*

What I read is one possible answer to the abstraction—intentionality, and quite frankly I don't completely understand what intentionality means, however, I interpret the Wikipedia definition to be about: (i) what is out there (*the ontological status of the entities*) and; (ii) what is in the mind (*objects of intentional states*); and (iii) how they are related (*intentionality*)? I use the word intentionality for what I believe it means, and what I believe is that it refers to what goes inside my head. Although my interpretation is correct or not, the only mind and its behavior that we can know is the one inside our head. I do not know if my view is in any way applicable to the word intentionality, but my use of blanket certainly describes a very interesting way for humans to exhibit sapience,

Intentionality is voluntary behavior initiated by the mind that is part of immaterial memory. It occurs anytime the mind decides to do something not prompted by external or internal stimuli. Once initiated, intentional behavior can be any one or combination of the basic modes of human behavior such as perceptual, procedural, or mental. It is a unique internal stimulus but is mostly conceptual. In addition to the spontaneous behavior—reflexive, perceptual, and procedural—that is common for all animals, we humans possess behavior that is based on our intentions exclusive of external and internal physical stimuli. As suggested, the sapient faculty {I} that resides in the immaterial memory initiates intentional behavior. The question we need to ask is not *why did {I} make that decision,*

133

but how? If {I} made the decision, then what is the referred to in the above example? Is there more than one source of decisions in the mind? I will answer this better in the next section *on thought*. We may not discover the true nature of mental decisions, but I am certain that neurons or neural maps do not make decisions!

In philosophy, intentionality is the power of minds and mental states to be about, to represent, or to stand for, things, properties and states of affairs. To say of an individual's mental states that they have intentionality is to say that they are mental representations or that they have contents.

With an immaterial memory, we can describe the mechanism in which the sensations entering the brain are transformed into electrical impulses that cause synapses to activate a specific signifier field just as electrons moving in a wire create a magnetic field. Each quale exists at a specific depth of the nous (q-map) in the qualia-field that is contiguous with the neural map in which several specific synapses are located. Each q-map is associated with a specific quale, for example, redness is experienced when a specific q-map is activated by a specific signifier field.

When I look at a red apple sitting on a table, my view of reality contends that the apple is nothing more than a spherical configuration of s-points from which photons or waves are reflected from the surface and by intromission the light wave (or photons) enter my eyes. The eye reacts with two distinct actions: (i) the action in which the retinal ganglia create, in the brain, the outline of the apple's shape and (ii) selective properties of the light wave cause certain neurons determined by the apple's shape to activate designated qualia (color) in the *nous*. This describes the "binding problem" mentioned earlier.

The main elements of the sapient faculty are mental representations generated by cognition, thinking, and thought functions. The mental function intentionality, in which the sapient faculty, determines whether a decision is to be made. The sapient faculty is involved with the initiation and control of both forms of behavior: psychical (thought), and physical (action). Intentionality initiates both physical and psychical behavior from within the mind. What about the mind allows us to think and to initiate physical behavior? Attempts to explain intentionality have resorted to some form of dualism such as process, substance, predicate dualism, or any of the other forms of dualism without success. The problem has been a reliance on abstraction, words that describe observation but do not explain. The question then is: how does the psychical activate the physical? I turn to the holonomic mechanism.

The nous that we have within our morphic s-frame (our body) is a particle of infinite nothingness called nous. The brain is the most complex material object in the universe, because of the material complexity, consciousness, and hence nous concentrates in the brain. Nous is a particle of the infinite nothingness that initiated the big bang and is creating the universe. The same impetus—infinite nothingness in the form of pneuma provides the impetus that increments the cosmic s-frame. The mind needs only to supply information, and the impetus to increment the cosmic s-frame that includes all the matter in the universe including our bodies causes whatever is experienced as motion. Because there are so many forms of thinking, for example, we might find ourselves: learning, memorizing, deciding, planning, contemplating, observing, disassembling, discerning, creating, discriminating, pondering, brooding, imagining, and many more, the best way to address the subject is to categorize them. There are three main categories: spontaneous, intentional, and internal. Free will provides choice, volition initiates action.

M. Personhood Path (vi)

Both the conceptual and perceptual paths arrive at the motivation faculty that determines the response to the input-information. There are three motivational levels each associated with a specific behavioral mode. At the lowest level, our *basic instincts* motivated us and other vertebrates, but only humans, because we are the only animals with minds and we are the only animals that can think rationally, possess two other levels of motivation: (i) tangential insight motivates rational behavior; and (ii) radial insight motivates transcendental behavior.

1. Motivation

The underlying reason a person acts. I identify three levels of motivation: basic instincts motivate animal behavior; tangential insight motivates rational behavior; and radial insight motivates spiritual behavior. This subject is discussed in Chapters 19, 20, and 21.

2. Personhood—{p}

Personhood is the psychical component of the soul that governs our behavior. To have a full understanding of the soul, we must examine the nature of human behavior, its stimuli, and its motivation. Human behavior reaches a level of complexity that exceeds anything else in the universe. The best we can do is to generalize normal behavior, and the only object that can be examined is one's own mind. Before addressing human behavior, I will take another look at consciousness, mind-science's greatest dilemma. However, if we accept the existence of nous, an immaterial substance, explaining psychical phenomena that science cannot explain, becomes more feasible. I described the soul in a way that can be useful as an effective argument. Based on the premise that God exists and the use of two modalities of space to describe the hylomorphic nature of reality; allows me to explain the consciousness, the major mind-body problem for mind-science.

The behavioral path leads to personhood, the culmination of our individual acts. Individual acts depend on all the elements of behavior, but primarily on motivation. the capability to decide when and how to act. Through thinking that involves intellect, reason, and one's values, in other words, with rationalization we modulate our basic instincts. personhood depends on cognition that includes motivation, free will, and volition. Whereas selfhood is innate, the development of personhood is ongoing throughout one's life. The constant thoughts flooding the mind every waking moment of every day remind us of the conflict between our personhood and our selfhood. What should I eat? What should I do? Should I read or watch TV? And, on and on and on. The answers to such mundane indecisive choices, most of which have inconsequential outcomes, illustrate the existence of a duality in the mind that is also present when the questions have consequential outcomes. It is this contest that determines the relationship of personhood and selfhood, the {I}, the integrity of the soul.

N. Soul

Only hard-core materialists deal with the nature of the soul directly by claiming the soul and psychical substances do not exist because they cannot be detected, a view from which science derives the imperious claim that only through science can the truth about the nature of reality be known. This application of scientism prevents science from explaining consciousness, which certainly seems to be immaterial. Consciousness that we experience is not material and it is not physical; hence it certainly must be immaterial and psychical.

Science's explanation of consciousness as an emergent property of the brain is an observation covered by an abstraction; it is not an explanation. If we accept consciousness as an immaterial, psychical emergent property of the brain, denial of the existence of a soul is inexplicable.

The soul exists in the immaterial memory; it is formed from two faculties: (i) selfhood (s) and (ii) personhood (p). Figure 10 shows the soul at the intersection of two mental sequences, one sequence is spatial, the other temporal. The *spatial sequence* of the soul derives from external proprioception, the sense as to where we are in the world, and that gives rise to selfhood. The temporal sequence of the soul derives from volition and the free will decisions between good and evil and between good and doing nothing that we make throughout our lives when confronted with a choice. It is such choices that determine our personhood. Although there is no general agreement among the great thinkers concerning the nature of the soul, there is a general acceptance that a soul exists and the best that can be said is that the soul is an internal faculty of living organisms composed of an unknown psychical substance.

However, because we accept the existence of consciousness, an immaterial, emergent property of the brain, it seems like pure dogmatism to believe the soul does not exist. My basic premise that, God exists, allows for an immaterial/ psychical substance called infinite nothingness that exists in three modalities: pneuma, bios, and nous. Then, the existence of the nous allows an explanation of subjective phenomena that science cannot explain and allows me to state that *consciousness is simply the manifestation of the nous.* So, if that is consciousness, what is the soul? Here is a selection of definitions:

a. My dictionary describes the soul as: "an entity that is regarded as being the immortal or psychical part of the person and, though having no physical or material reality, is credited with the functions of thinking and willing, hence, determining all behavior."

b. In Sacred Scripture, the term soul often refers to: (i) human life; (ii) the entire human person; (iii) that which is of value (iv) that by which we are in God's image.

c. A philosophical view defines the soul as: (i) the internal principle by which we think, feel, and will and by which our bodies are animated; (ii) an incomplete substance which is simple and connaturally related to the body but intrinsically independent; (iii) infused in the embryo at conception and is immaterial, immortal, individual, and formative while forming the body.

d. The theological view is that the soul is: (i) human life or the entire human person; (ii) the innermost order of man; (iii) that which is of great value in him; (iv) that he is most in God's image; and (v) the psychical faculty in man.

e. The Italian philosopher priest Thomas Aquinas (1225-1274) added that the rational soul is: (i) with the sensitive, vegetative principle, the form of the body; (ii) an incomplete substance because it has a natural aptitude and exigency for existence in the body; (iii) in conjunction with the body makes up the unity of human nature; and (iv) not wholly immersed in matter, its higher operations being intrinsically independent of the organism.

f. To complicate matters, we have Aristotle's *vegetative soul,* the *sensitive soul,* and the *rational soul.*

I describe the soul in a way that can be useful as an effective polemic. Here is what I believe: the soul is not material; it is immaterial. It obviously is not constructed from neurons but in some way depends on them. Based on the premise that God exists, I earlier used two modalities of space to describe the hylomorphic nature of reality; this allows me to define consciousness as the manifestation of a psychical substance called nous, a modality of the infinite nothingness. From there I postulate that the soul is the internal form of the nous, the substance of the immaterial memory. The soul is formed as the difference between the selfhood {s} and personhood {p}. Selfhood is our basic animal nature, our being; personhood is our becoming, our success in advancing along the path of righteousness. Selfhood is what we are; personhood is what we become. The choices we make throughout life create our personhood that often conceals our selfhood, the faculty hidden in the world, the one that we cannot fake, but can bury. {I}, the faculty of the mind that makes decisions and forms the integrity of our souls, the faculty that appears as one's personality modulates the ongoing conflict between personhood and selfhood, The real me, selfhood, can only be known to God. What the world sees is our personality, but that is mostly the manifestation of the personhood's success in modulating selfhood. Personality is the observed relationship between personhood and selfhood. In general, in most people, personhood dominates. However, experiences such as frustration or a perceived insult can expose the selfhood of an otherwise pleasing personality. Only the true *reasonable adventurer* has a sufficiently elevated level of personhood to constantly suppress selfhood.

Chapter 19 —Rational Behavior

Rational behavior falls within the social norm for optimal benefit. Each member of a social unit is expected to conform to the social norms consisting of the laws, rules, and mores. The social norms are those which we humans have decided, as a consensus of our social unit, is beneficial for both the individual and the social unit. Social units are as small as the family or as large as a race, with other levels in between. The norms are set as limits on behavior to keep peace because humans (especially males) can fail to control their basic instincts resulting in irrational behavior. Whereas the social norm applies to the social unit, the optimal benefit applies to the individual. In other words, rational behavior means each person behaves for their own benefit (freedom) without violating the social norms (liberty). Furthermore, before we can examine irrational behavior, we must first understand the rational by examining the changes that elevate humans to a higher level of behavior. Rational behavior is restricted to humans because only we humans have minds, a novel faculty among the animal-kingdom

The major difference between human and non-human animals is the presence of the mind. The mind, a faculty that only humans possess, is a combination of a language faculty plus nous. Since metazoans are sentient, they possess nous, so the main addition to the human corpus is the language faculty. The mind's ability to create, differentiate, synthesize, and attach meanings to words has raised the level of human mental capabilities far above the rest of the animal kingdom. The mind adds mental elements to the human corpus that no other animal possesses.

When the first humans acquired language, it created the mind that possessed a powerful mental faculty, the *sapient faculty*—symbolized by {I}, and we acquired the capability of controlling our basic instincts. {I} has two components (p) and (s) standing for personhood and selfhood. Personhood is determined by how we apply free will when confronted with a choice between a good or a bad response.

An important increase in behavioral complexity in humans is due to the addition of a second level of motivation, called tangential insight, the ability to think before acting; it is unique to humans. Tangential insight consists of three factors: *intellect, reason*, and *values*. Tangential insight motivates behavior by suppressing or enhancing animal behavior. Motivation of basic insight usually induces rational behavior; without rational motivation, basic insight can lead to or even enhance irrational behavior. In addition to tangential insight, symbols provide a new form of stimulus. Our animal body becomes a *corpus* with the addition of the mind. I represent the human mind to be free as depicted by the symbol {I} shown in the immaterial memory. The {I} allows humans to think and act intentionally. Since we humans are animals, our basic instincts motivate most of our behavior, but the mind adds complexity to our perceptual and procedural behavior and adds capability for thought and the burden of irrational behavior.

Humans are capable of procedural, perceptual, and intentional behavior. As animals, humans are also capable of reflexive, procedural, perceptual spontaneous behavior, and not shown, spontaneous mental behavior (thought). Reflexive behavior is always spontaneous.

There is no evidence for my explicative level, but I contend that, if God exists—and that is my basic premise—then there is a psychical component in the body (nous) in which the sapient faculty induces the morphonomic mechanism that supplies the information that reconfigures the morphic s-frame to behave in a way that achieves a selected objective.

I contend that the only way to describe (forget explaining it) human behavior is to postulate that God exists and assume that a psychical—or spiritual if you prefer—substance (nous) exists in our cranium along with the brain. Like the brain, the nous is necessary but not sufficient but only with the presence of the nous is the brain sufficient for thought to occur. Science has trouble accepting the necessity of a psychical substance because it is not observable or measurable. Even though dark matter and dark energy are not observable, science is willing to invent them because they help to describe certain observations.

Because of language and the ability to think, we humans possess a mindboggling range of behavior, from the lowest animalistic to the transcendent. When God created the first human by creating the language faculty, and with it a mind, a whole new set of experiences became possible to create rational reality. The mind acts as an arbiter of stimuli, filters the stimuli through the motivating elements, and chooses how to behave. Rational behavior when initiated by sensations, symbols, or {I}, can result in reflexive, perceptual, procedural, or mental (thought) behavior or a combination of the four. Describing human behavior precisely is difficult because in one day, or one hour or less, we might behave in a single mode or in combinations. The best that we can do is to proceed cautiously with generalities. I choose to categorize rational behavior by the nature of its stimulus as spontaneous or intentional.

At the explicative level, the corpus of an animal is an autonomous, hierarchal configuration of s-points imbedded in the cosmic s-frame (we are truly part of the whole). When the {I} of a human chooses an objective, its motion is the result of an incremental series of reconfigurations (n's) determined by the pertinent algorithm.

Algorithms stored in immaterial memory are the driving force of both growth and movement of a human. In time, the morphic s-frame follows a path of incrementation modulated by a morphonomic mechanism stored as concepts in immaterial memory until the human reaches' maturity. With learned behavior, a corpus' increments through a sequence of reconfigurations (morphic s-frames) that are determined in accordance with a *meaning* implanted from without. The degree to which the reconfiguration occurs morphonomously is a function of where the corpus stands on the evolutionary scale. In humans, reconfiguration choice is complete; the nous makes the determination with complete autonomy despite instinct or conditioning. The mental process manifests as the thing we call "the will."

The explicative level of an efferent event happens, for example, when the {I} decides to smile, and the analog nous causes the appropriate digital neuron to activate our facial muscles following a series of neural connections that was innate or learned motor memory. Without the presence of the psychical substance, science contends that a neuron or neurons make the decision and spontaneously activate a smile. The question arises as how can an organization of hydrogen and carbon atoms make decisions? My view is they can make the smile but not the decision. Figure 11 depicts a model of the rational behavioral elements; it includes all the animal elements from animal behavior (figure 6) plus the following changes: (i) Corpus adds language, mind, and the sapient faculty {I}; (ii) Stimuli adds symbols; (iii) Motivation adds tangential insight; and (iv) Behavior adds cognition and intentional behavior.

A. Rational Corpus

Humans are animals, have physical corpuses, but differ from all other animals psychically because we have minds. We have minds because our brain, that part of the corpus, contains an arrangement of neurons that form a language faculty that forms words with which we think and bracket percepts.

B. Stimuli

Although stimulus only applies to the initiation of the behavior, I will continue to use it as a categorization factor because psychical, immaterial memory behavior is mostly intentional; and physical material behavior is mostly spontaneous. Reflexive and procedural behavior occur in the material memory and hence are physical; perceptual and mental occur in the immaterial memory and hence are psychical. However, no matter how we categorize and focus on modalities, there is still some overlapping and multiple *symbols* are signs with meaning; they arrive externally primarily as words, both orally and visually. They have specific meanings that induce thought and action or storage in the immaterial memory as knowledge. Language, in this thesis, refers to symbols stored in the material memory as neural maps and are associated with meanings and percepts stored in the immaterial memory. Symbols create thoughts that allow the mind to reference percepts with words. Symbols allow humans a greater range of behavior. Animals sense their environment and extract narrow specific meanings and must react in extremely limited ways. Humans are animals and often react to signals but can choose to react against basic instincts because of free will. Will is free because the human mind can apply insight to provide alternative reactions to varying degrees. Language, composed of innumerable combinations (words mainly) of discrete units (letters), endowed humans with an opportunity for creating meanings. Words allow humans to think internally and behave rationally. Words are not the only form of symbols, for example, there are: numbers, Morse-code, musical notation, flags, electrical + and —, maps, diagrams, and charts. These are visual inputs that have specific meaning. Symbols are mostly digital (discrete). In addition to symbols, humans added a more important stimulus, intentionality.

Note 35: In my view, both signals and symbols are signs with meanings. The difference is that signals are analog (continuous); symbols are digital (discrete).

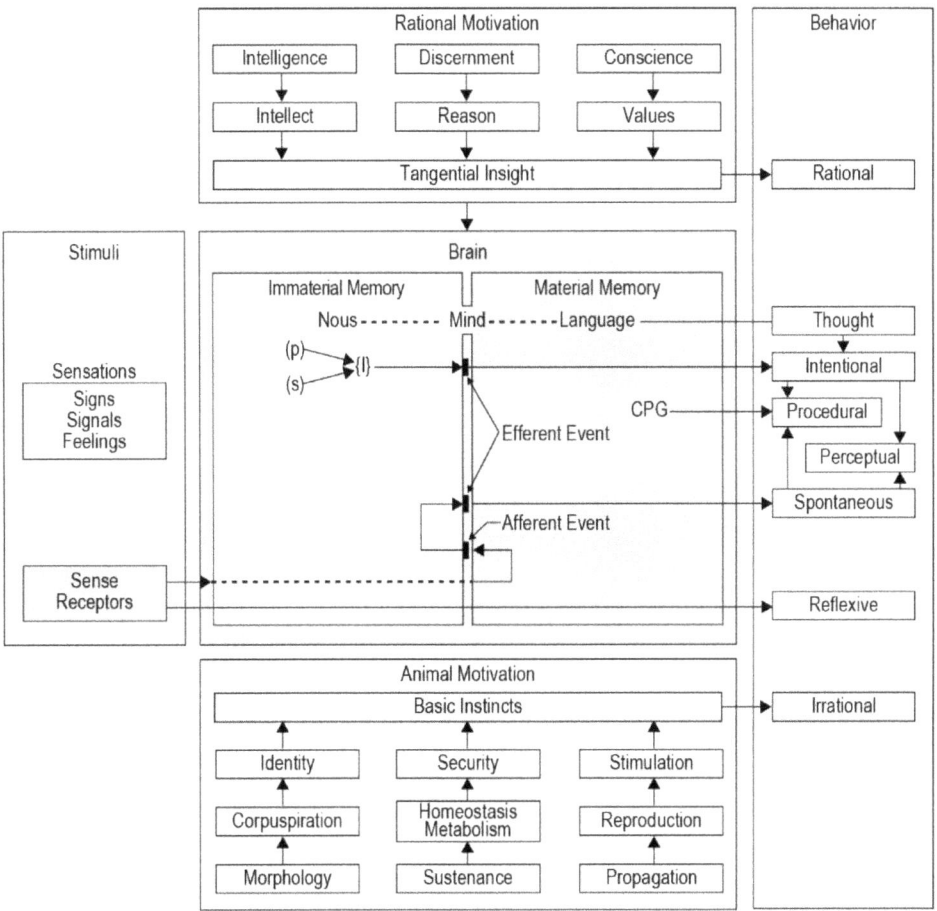

Figure 11. Rational Behavior

C. Rational Motivation—Tangential Insight

With tangential insight as the motivation for behavior, normal people behave rationally or at least do no harm. Instead of a single level of motivation found in animals, there are three levels of motivation available to humans: basic instinct, tangential insight, and as I will explain in the next chapter, radial insight. Human behavior, then, is based on the interaction of three levels of motivation: basic instinct inheres in the body; tangential insight inheres in the mind; and radial insight inheres in the soul. Without positive modulation of tangential insight, our motivation can resort to the basic animal instincts and a possible resulting irrational behavior. However, believing that tangential insight is a filter that always blocks instinctual behavior is foolish, because the mind can apply free will to ignore insight. A complex interaction of instinct and insight, which most of the time is unconsciously spontaneous, determines the direction in which motivation tends.

As basic instincts are the foundation of selfhood; insight is the foundation of personhood. As stated previously, intellect depends on one's intelligence, the ability to acquire, consolidate, and retrieve knowledge. The basic contribution of intellect to insight is that it is essential for developing the ability to reason and adopt a healthy value system. Intellect without reason and values can do more harm than good.

1. Intellect

Intellect is simply the knowledge that is stored in immaterial memory. We can assume that intellect is not a direct motivation; its main value is broadening a person's ability to choose one's behavior through reason. We use intellect to develop the ability to reason and to develop a positive value system. Intellect alone is a poor modulator of basic instincts. In fact, it is the element that can enhance instincts in the negative direction. Low intellect is one cause of failure to prevent irrational behavior because it is often the failure to develop a virtuous value system. However, mostly the type of irrational behavior resulting from low intellect is harmless, especially when compared to high intellect people that use their gift to enhance irrational or even evil behavior. There are examples of high intellect persons that are extremely evil, Hitler for example.

2. Reason

Reason is the faculty for thinking coherently, logically, and with sound judgement is called reason; it is a mode of thinking that is based on a spectrum of capabilities ranging from common sense to logic. Reason appears in science as plausibility. To informally apply reason to our acquired knowledge means to form sound judgements based on an innate sense devoid of contradiction, hypocrisy, deception, exaggeration, prejudice and incorrection. To reason formally simply means to apply the principles of logic to one's judgements. Unfortunately, we cannot apply formal logic effectively to most judgements that we make. Informal reasoning occurs within the limits of one's intellect and values. If we do not include toleration of the value system, then reasonable moderation of the identity instinct is more difficult, and we consider our identity to be superior to another's. Religion without the application of reason can lead to intolerance. Fortunately, the gift of language, the presence of a mind, and the ability to think that it entails, adds a capability for moderating our animal motivations to provide an additional mode to the human behavior called rational behavior. Unfortunately, the ability to think also provides a capability for rationalizing undesirable animal behavior leading to irrational behavior. This dual option demands choices for humans not present in other animals. This is free will.

3. Values

The value system is the mental faculty that makes the decisions concerning the development of one's personhood. It is the intentionality, motivation, and free will part of the personhood path. It deals with the social, moral, and personal ethical principles that influence our conduct that transcends one's own self-interest. The value system is only one part of tangential insight and tangential insight is only one part, along with basic instinct and radial insight of motivation of any single act. value is the faculty that decides the integrity of one's soul. Although our conscience provides an innate sense of values—such as right and wrong, rational, and irrational, true of false, good, and evil, justice and injustice, and other moral dichotomies—we acquire most values through inheritance, origin, and education.

Inheritance that we are born not only with an innate conscience, but also with both selfhood and the capability to form our personhood. The difference between selfhood and personhood is an expression of the integrity of the soul, one's personality. My wife and I have 8 children, 22 grandchildren, and presently have 19 great grandchildren and every one of them was born with a unique personality. In my view DNA determines selfhood, and traducianism determines an incipient personhood in each one of us. In addition to DNA and traducianism, my view of a parallel world suggests that we are born with an evolving conscience from a previous life. A conscience is the innate function associated with the value system; it is associated with the volition faculty in immaterial memory.

Origin, one's place of birth and other factors such as physical characteristics, and health affect the development of a value system. People are born into freedom or slavery; rich or poor; with or without good appearance; in good or poor health. And these are factors that determine the responsibility, difficulty, and capability for developing a desirable value system.

Education is the primary source of a value system that is gained mostly by one's family. Parenting has the greatest influence on one's value system, especially when religious factors are present. Parents that have healthy value systems pass them on to their children, especially if the parental influence derives from a religious environment. But parenting and religion are no guarantee, and the absence of value-based parenting and religious environment does not necessarily prevent development of a positive value system. The innate component of the value system can be sufficiently positive to allow people to overcome deficiencies that result from a poorly learned component. The innate component appears as conscience, which can be overly scrupulous, healthy, or completely deficient. The development of a healthy insight has a much higher probability of success when children grow up in a normal family environment.

D. Modes of Behavior

We humans have a far greater range of behavior than the rest of the animal world which are limited to three spontaneous modes listed first in the following list.:

1. Spontaneous Rational Behavior

Spontaneous behavior, our most basic behavior, is stimulated by our sensations, and takes place through elements of our material memory. Hence, spontaneous behavior is mostly physical. Spontaneous behavior appeared previously in connection with non-human animal behavior. As indicated in figure-11 the Rational Behavior chart, spontaneous motivation is mostly associated with the three physical behaviors: reflexive, perceptual, and procedural modalities. However, in humans, there is also spontaneous transcendental behavior.

2. Spontaneous Reflexive Behavior

Reaction human behavior, made possible by the presence of the mind, begins at the rational level of motivation, but because humans are animals, I discuss the physical modes of behavior: reflexive, perceptual, and procedural. In fact, the basic instincts that motivate animal behavior are the source of some rational behavior and all irrational behavior. In humans it is the same as in other mammals, we react to certain signals without thinking. Hence, we sneeze, and we scratch where it itches, we duck to avoid an object aimed at our head, and we step on the brake pedal at red lights. However, humans can apply free will to the stimuli that normally calls for reaction and we stifle a sneeze, suffer without scratching an itch, block the object with our hands, and run the red light. Humans learn many activities calling for reflexive action. The most difficult activity requiring reflexive skill is hitting a baseball. A 90-mile an hour fastball reaches the plate in about 0.4 seconds (400 milliseconds). It takes about 100ms for the brain to process the trajectory and 150ms to swing the bat to a position in which the bat makes contact. There is less than 10ms to swing the bat in the position of hitting the ball into fair territory. This decision to swing certainly happens without thinking and is surely reflexive. It is surprising that professional ball players are successful in making contact greater than 50% of the time they swing. Since it takes us 3–400ms to blink our eyes, something else besides normal muscular reflex action is involved. Just what that might be I will discuss later. Reflexive behavior can be innate like sneezing, or we learn it, like hitting a baseball. A sense-receptor does not stimulate hitting a baseball, it is the perceptual intellect in the immaterial memory that initiates it.

3. Spontaneous Perceptual Behavior—Observation

Focusing, watching, looking, noticing, inspecting, viewing, listening, sniffing, touching, and sensing are ways that we observe. They are modes of perceptual behavior. Perceptual behavior in humans is like that in other animals; we focus our attention on our proximate environment while our action focuses on a specific objective. The scene of our activity (a percept) becomes the stimulus for behavior. However, perceptual behavior in humans differs from other animals in its diversity of options. We act perceptually when we converse, paint, play sports, drive a car (at least we should focus), garden, and any other activity requiring focus, with a minimum of thought, on specific perceptual inputs. Unless we are stationary and studying a scenario like painting in a museum, perceptual behavior mostly includes motion of one type or another. Perceptual behavior is physical in that the stimuli are sensations. Because it requires minimal thinking, perceptual behavior gives the mind a rest. I argue that the amount of memory that a normal mind contains must exceed the capability of material memory. Given my own experience, I contend that it is immaterial memory in which the percepts reside. We cannot recall the percepts that are without a signifier, we can only recognize them. Hence, baseball players can successfully catch a fly ball without thinking are recognizing a signifier because they can judge the trajectory of the ball. And what is true of fielding a baseball is true in all other perceptual behavior, it is most successful when devoid of thought. For example, the difference in hitting a baseball (reflexive) and catching a baseball (perceptual) is timing. And like most modes of perceptual behavior, a person's proficiency in any specific perceptual activity is an inheritance (perceptual intelligence) that person has honed with intense learning.

What I have just described is perceptual behavior of which we are capable just like the rest of the mammals, but we possess a mind that they do not, and consequently we have a far greater spectrum of perceptual behavior than the rest of the animal world. Pottery, painting, gardening, sports, most blue-collar jobs, and I believe, music, and other activities in which we engage happen with perceptual intellect. Perceptual behavior is more temporal than spatial, and proprioception allows a person to alter one's motor memory with feedback from the changes in the perceptual input. Just as a cheetah chasing an antelope will alter its path to conform to the antelope's, a jazz musician can intentionally, and without conscious thought, improvise and deviate from the score. However, the cheetah is behaving perceptually, as is the musician, but the musician is also behaving procedurally.

4. Spontaneous Mental Behavior—Reverie

Not all thought is intentional; focused thought does not consume all the mind's time, reflexive and perceptual behavior make their demands as does, what might be the biggest demand of all, daydreaming (reverie). Reverie is spontaneous psychical thought of an unfocused mind. A reverie is the flow of random, pleasant, or worse, innocuous thoughts that may not start out randomly, but soon can lead us down completely unpredictable paths. Sometimes referred to as "stream of consciousness," a term that surely resonates differently with each mind. I am not sure what it means because like everyone else in this world each of us can only know one's own mind. In my case, reverie is the work of the {I}, the sapient faculty that controls all thoughts and if that is true, I must deal with an {I} that is suffering from a case of overactivity for he takes me down very divergent paths. Whatever the source of reverie, it surely inheres in the immaterial and within {I}; our selfhood may have two sides. Or there is another mental faculty present in the immaterial memory that moderates the sapient faculty; it is the soul.

When we silently manipulate (mime) words, we are engaged in two mental functions, memorizing, and recalling. What that involves is the conscious interchange of words in the material memory with their word images in the immaterial memory. We mouth words, and we think with word images that are churning around in the mind and are not conscious unless interrupted by {I}.

5. Intentional Rational Behavior

Behavior stimulated by our sapient faculty, {I}, separates humans from the rest of the animal kingdom. {I} is the core faculty of intentional behavior. Rational behavior is physical behavior when associated with material memory, or transcendental behavior when associated with immaterial memory. Intentional rational behavior has four components: (i) thought, (ii) perceptual and procedural behavior, (iii) learning, and (iv) memorization.

Thought is completely personal; hence, we can only guess at others are thinking based on their behavior. Psychologists do their best to extract a patient's thoughts and can with confidence make deductions based on generalizations of observed thought patterns. Still, no one can ever know what is in another's mind, hence, what I write about thinking is based on my own thoughts, which judged by my normal history of behavior, are recognizable by a large contingent of similar thinking minds. Thinking is a mental activity that involves digital symbols, primarily words. The material memory (the brain) stores the digital symbols in the neural maps.

When we think about thinking we soon end up with a long list of words that apply. There are ways to categorize the various modes of thought. The simplest way is to group them by whether thinking is intentional or spontaneous. Because always focused, Intentional thinking, that which begins with {I}, invariably focuses on an objective. When learning, deciding, pondering, wondering, rationalizing, reasoning, worrying, contemplating, concentrating, meditating, imagining, and any other focused mental exercises we engage in, we do it to achieve a specific goal. When we are concentrating, we are concentrating on something. When we rationalize, we are justifying something. This is what I define as thinking, including negative thinking such as when agitated, disappointed, aggravated, our thinking focuses on the source of our negativity.

When we speak, we think about what we say. Although there are times when we speak without thinking; we can speak reflexively, and we can speak perceptually. Speaking, since it involves facial muscles, is a physical mode of behavior. Reading is an active mode of thinking because what we read is being set in our short-term memory, and if your intelligence is efficient, that contained and discerned in the short-term memory transfers into the explicit semantic memory.

6. Intentional Procedural Behavior—Action

Action (physical behavior) takes care of the unconscious procedural movements that are essential to survival such as walking; they are innate with learning applied to initiate or upgrade them. In addition, we devise innumerable activities that are not essential to survival that we learn, such as reading, writing, playing musical instruments, throwing a ball, dancing, and other activities too numerable to list.

Most non-essential intentional procedural behavior is both innate and learned. The innate part is talent; the learned part is skill, especially those activities that are common to humans such as playing musical instruments. Musicians achieve competence by learning musical notation and constant practice. Professional musicians have stored in memory enormous amounts of musical information. It is amazing to watch a classical pianist play a composition like the 3rd movement of Beethoven's Moonlight Sonata then realize that she has stored more than one composition in memory. Such talent must require more than what is possible in the neural maps in the material memory. At the explicative level, there are two memories to deal with, so we can store the digital part, the notes, in the material memory and the analog part, the melody, in the immaterial memory. My only musical experience is with singing (poorly), and I know far more melodies than lyrics. Nevertheless, it is easier to remember words to songs than to remember words of a poem; hence the analog guides the digital, that the motor memory guiding a pianist's hands and fingers, the digital, to the proper keys is set in the pattern of the music stored in the analog immaterial memory. If music were simply the result of the memorized memory in neural circuitry, pianists would be without their ability to interpret. Interpretation resides in the perceptual memory and is a consequence of the perceptual intellect of the pianist who in transforming the procedural to the perceptual and the rational to the transcendent actualizes beauty. I cannot imagine a composer writing notes without mentally humming the melody.

Music is both conceptual and perceptual because the score is digital knowledge stored in the material memory, and the melody is perceptual intellect stored in the immaterial memory. For a singer, the lyrics (words) serve the same function as the score (notes) in both cases of which the melody modulates the song. It is analog music (the sound) that achieves transcendence in the creation of the beautiful, and the experience of beauty, without which music could never achieve transcendence in the form of beauty. A musician plays and converts scores of digital notes into an analog melody. The material memory contains the score, and the immaterial memory contains the melody, The musician or singer then allows melody to control the score, and the music seems to flow. There seems to be a faculty in immaterial memory that does the same job in remembering sequential information that the central pattern generator (CPG) does in material memory.

Note 36: The word 'central' does not mean that there is one CPG, there are CPGs distributed throughout the nervous system).

The mode of behavior associated with our animal nature such as walking, running, swimming, chewing, and any other sequential movement of the body or its parts. These movements are based on motor memory. We learn to walk, run, and jump, but the basic motor memory, which consists of neural channels needed to perform those acts, is innate; the program for accomplishing a specific procedure such as walking becomes stored in the

146

material memory as neural channels. However. It is intentional procedural behavior that separates us from the rest of the animal kingdom. Procedural behavior such as dancing, singing, athletics, and learning poems and music compositions requires practice. Repetition improves motor memory and its performance. The key neural element that is essential to procedural behavior is a central pattern generator.

CPG's generate synchronized and rhythmic impulses that then synchronize the movements. Most simple procedural behaviors are either innate like swallowing or quickly learned like walking. However, we humans have invented a plethora of procedural behaviors like playing musical instruments or singing and dancing, that require a different form of memorization than motor memory. It is perceptual procedural behavior because both modes of behavior are essential. Something in the immaterial guides a pianist's fingers in a remembered sequence. It cannot be neural channels that map a song, because good musicians can improvise by leaving the remembered sequence to eliminate or add notes to the notated sequence.

7. Intentional Perceptual Behavior—Focus

Intentionality exists in immaterial memory. Perceptual behavior is mostly spontaneous because the sense receptors initiated it. All animals, including us humans, respond to sensations. Hence, we answer the doorbell when it rings (a signal), go to the kitchen when we feel hungry (an appetite). Those are spontaneous. Perceptual behavior can also be intentional. Humans respond perceptually when {I} decide to indulge in certain thinking free activities such as gardening, knitting, and driving a car.

High *temporal perceptual intellect (TPI)* is an important trait to have for dancing, driving, playing sports, and playing musical instruments. The ability to discern the changing scene, an ability that I believe is innate, is high, for example, among the best hockey, soccer, and basketball players, TPI is a function of the proprioception sense. Knowing where you are in relation to the other players, both teammates and opposition players, is a great advantage for being successful. However, to be a great hockey player one must learn not just how to skate but how to skate skillfully.

8. Intentional Mental Behavior—Learning

In evaluating the collective capability of the billions of individual minds one finds a tremendous range in both depth and breadth. Virtuosity in music, mastery in art, technological expertise, scientific discoveries, invention, diplomacy, revelation, and all other human activities that contribute to the increasing knowledge of the world are the result of learning, that special mode of thought.

Learning is the input of signifiers that form neural maps in the material memory and simultaneously form signifier images in the immaterial memory. The short-term memory inheres in the neural maps of the material memory. Long-term memory resides in immaterial memory. Learning then is the process of acquiring information, consisting of a signifier, attaching it to its referent, and storing the signifier image and referent in immaterial memory.

Learning is one of the most important forms of mental behavior. We are born with all the faculties and functions of immaterial memory; but we developed it through what we learned in life. We acquire knowledge by learning, and we learn through instruction, reading, study, observation, and other avenues of acquiring information. Learning focuses our attention on something that is unknown to us. The mind, the interface between material and immaterial memory, is the learning faculty. The knowledge that the mind acquires through learning is both conceptual and perceptual. Conceptual learning depends primarily on sight and sound, mostly through reading, seeing, and listening. Perceptual learning depends on all the senses, sight, hearing, smell, taste, touch, proprioception and on the acquisition of facts and other pertinent information relative to a perceptual signifier.

Digital information includes words, numbers, symbols, signals, and musical notes that have been formed as ideas, equations, diagrams, and musical scores that have been created in an individual mind and then are: (i) instantiated as c-noumena stored in the realm of potentiality as rational reality, and (ii) actualized as part of the general digital information stored in books, and electronically—for example the internet— and other forms of digital storage that forms the noosphere. The main sources of conceptual knowledge are mathematics, physical science, and fact-based subjects such as history.

Conceptual learning—we acquire it primarily through reading. The higher the intelligence, the less learning required to store a specific particle of knowledge. Highly intelligent people store knowledge with a minimum effort. Our efficiency in acquiring knowledge depends on our innate ability to comprehend concepts and our efficiency in storing it in long term memory. However, studiousness in learning can compensate for intelligence. This depends on the effort required to transfer knowledge from short to long term memory. In my view, what I am describing is the efficiency of the afferent event regarding its ability to hold information in short term memory, the duration of which determines the probability of retaining the information in long term memory. Currently the IQ tests we take measure conceptual intelligence, the capability of dealing with symbols and their relationships. People with high conceptual intelligence tend to become scientists, mathematicians, and philosophers.

Study is the primary method for acquiring conceptual knowledge, which depends on one's intelligence and perseverance. Intelligence is innate; we acquire perseverance; and it supplements intelligence. Intelligence is the capability for learning and learning is the capability of committing the results to memory. However, efficient study supplements what a mind lacks in intelligence.

Perceptual learning, I believe, is innate. It is the capability of remembering perceptual information, the facts and knowledge associated with individual percepts. People with high perceptual intelligence are adept at remembering faces, names, facts, places, and scenes. Persons with high spatial perceptual intelligence tend to become artists, poets, lyricists, thespians, and any other activity requiring talent. Also, perceptual behavior has resulted in scientific discoveries, including Einstein's—the greatest of all. His perception drove his theories; perceptions of travelling alongside a light wave; a clock tower; a transverse light beam in a moving elevator.

Observation is the second form of learning, especially regarding perceptual intelligence. Perceptual intelligence can be associated with a specific sense. Thus, we not only have musicians and artists, who have a heightened sense of sound and vision, but there are also people with equally heightened sense of taste and smell.

Procedural behavior is that which is hard-wired in the neural map in the material memory, and specific stimuli produces specific behavior. For example, if you touch a hot surface, you spontaneously remove your hand. The neural map that activates the muscles that withdraw the hand are innate. We innately respond to loud noises, obnoxious odors, hunger, and dangerous sights. All reflexive behavior is spontaneous, and innate. These modes of behavior arise at the animal level of behavior motivated by basic instincts. Rational and sapient behavior are *psychical* because they are motivated by tangential insight and arise in immaterial memory. Transcendent behavior is psychical because it is motivated by radial insight.

Memorization functions include implicit, episodic semantic explicit, sensory, short-term, and long-term memory. Sensory memory retains input information, both perceptual and conceptual, for a brief period and is not part of the learning process, it merely acts as a gate that selects what enters the short-term memory. When the short-term memory receives a selected word, phrase, or another type of signifier, it incorporates in the material memory as a neural map, then a percept, consisting of the signifier image plus a referent, incorporates in the immaterial memory. The incorporation efficiency of storing the percept in the immaterial memory is acquisition part of one's intelligence; it decreases with age. The imprint efficiency is innate, at its highest as a toddler with a long slightly decreasing slope and a sharply decreasing slope in old age. Imprint efficiency depends on two factors: (i) the innate efficiency, and (ii) the duration of *mirrored* imaging, i.e., duration of the in short-term memory's signifier engagement with a signifier-image within the long-term immaterial memory. Acquiring knowledge consists of two channels; the referent of a percept incorporates directly; the signifier image requires learning. Perceptual knowledge and the efficient memorization of perceptual knowledge derives from our animal nature.

We can translate my view into the common view of the memory functions by recognizing that the material memory translates as the implicit memory, and percepts incorporated in the immaterial memory translate as the episodic function of the explicit memory, and concepts incorporated in the immaterial memory translates as the semantic function of the explicit memory. Subjective experiences cannot be translated because they are properties only of immaterial memory that do not exist in the common view.

Most perceptual behavior occurs spontaneously; as we increment through our environment, the mind acquires percepts incorporated in the immaterial memory. We notice the sights, sounds, odors, and react to them as animals do, primarily we recognize whether the environment is a familiar or a different perceptual experience. Immaterial memory acquires new scenarios for evaluation and ignored or sent to short-term memory for incorporation into long-term memory. When something in the scene appears noteworthy, a signifier (a name) incorporates in long-term material memory while the details of the scene become a percept (a referent) stored in the immaterial memory. All but the most dramatic scenes may require repetitions for incorporation in long-term memory. When revisiting an incorporated percept and we recall the signifier, we may also recognize the scene (percept) may or recalled.

The ability to retain percepts is a measure of one's perceptual intellect. Like conceptual intellect, where there is a tradeoff between the intelligence and depth of study; perceptual intellect depends on the tradeoff between perceptual intelligence and observation. How well each percept and signifier reside in long term memory depends on how often we encounter the specific scene. The mind often contains an enormous amount of perceptual information, and when we re-encountered previously experienced perceptual scenes the mind prompts the thought, "I have been here before!" Concepts, on the other hand, require more effort or study before recollection.

9. Irrational Behavior

Our behavior is rational when our actions fit the social norm for optimal benefit. Irrational behavior occurs when we act outside the social norm whether it is for optimal benefit or not. However, irrational behavior is not necessarily bad; there can be irrational acts of bravery.

Since my view is based on the premise that God exists, I must address the idea that God can be involved with how humans behave. God's involvement raises the basic question asked by the cynic: if God is all good why does God allow evil to exist? This is one of those questions a believer must deal with when arguing with a non-believer. In arguing for the existence of God—the main purpose of my thesis—we theists invariably go to a "God of the gaps" type argument and contend that "God does not create evil, people do." That approach, understandably, carries little forensic weight. My approach is to find a way to describe how God is satisfied in leaving to humans the job of dealing with evil. I have, but first I need to establish a more solid footing for my argument. That requires an examination of irrational behavior because although all evil behavior is irrational, the contrary is not necessarily true, and not all irrational behavior is evil. Irrational evil can be good or bad, for example when a firefighter enters a burning building, an irrational act, but if to save a child, the irrational becomes the heroic. Obviously, irrational behavior spans a spectrum of behavior from heroic to evil. So, why irrational behavior, why evil?

The survival of the fittest, strongly oriented towards self-preservation, creates a brutal way of life. Consequently, human progenitors survived, and we inherited, in our genes, like other animal species, a powerful set of instincts to fulfill the need for identity, sustenance, and stimulation. Since survival emanates from man's animal nature, it is understandable that basic instincts play a big part in our irrational behavior. Irrational behavior does not exist (however brutal and irrational it may seem) in other species, because basic instincts alone motivate behavior.

God designed us with a faculty for language, a mind, and the ability to think to develop insight with which to modulate the basic animal instincts and to find the most rational way to act that at times requires suppression of those instincts. However, the mind also has the power, because of free will, to use rationalization to suppress our insight and enhance our basic animal instincts in an irrational direction.

There is an excuse for humans to suppress insight to enhance the motivational power of the basic instincts. Fulfilling the basic needs results in subjective feelings that are superficial but enjoyable. We enhance our need for identity and experience pride, fame, and/or power that often motivates irrational behavior as brutal as murder. We enhance our need for security, and we experience pleasure and gratification that often motivate irrational behavior such as gluttony and addiction. We enhance our need for stimulation, and we experience satisfaction, excitement and thrills that often motivate lust, conflict, and perversion. God has provided us with motivational incentives to induce us to fulfill the basic needs and hence we can experience such motivating factors that we recognize as pride, pleasure, and satisfaction. However, if our insight is not sufficiently effective, especially when it lacks a positive value system, the mild inducements to live, to sustain life, and to propagate, become temptations and the power of our mind can drive those same beneficial temptations to motivate extremely irrational behavior such as murder, addiction, or perversion. Except for the true ascetic who has the willpower to suppress even the most powerful demand of their instincts, we, at one time or another have ignored insight and surrendered to temptation. I, of course, can know only one composite of these subjective experiences, my own. To complexify further the problem of explaining behavior, all six of the rational motivational factors can vary in strength that depends on their specific innateness and conditioning.

Like rational behavior, our irrational behavior is spontaneous or intentional. God created humans with insight to modulate the basic instincts which if not modulated can lead to irrational behavior that spans a spectrum of irrationality from overeating to committing horrendous crimes. But God did not provide humans with perfect insight that would prevent the basic instincts from motivating evil behavior. Far too many people possess tangential insight weakened by deficient intellect, inability to reason, or an absent or distorted value system. Failure of tangential insight to modulate our basic instincts causes not only evil, but also other kinds of irrational behavior.

We react to a stimulus, based on motivation by one or more of the instincts. Usually, when identity is the motivation, behavior is outward and directed towards others. Consequently, the need for identity is the primary source of evil. The need for security and/or stimulation tends toward personal irrationality. The need for security and stimulation are transient needs that flow through peaks and valleys; the need for identity is persistent, always present, and easily hidden within one's selfhood under a mask of personality. Whereas security and stimulation instincts are physical and over which we do not have complete control, the identity instinct is psychical for which there is complete control. The identity instinct is who we are; our physical self (our body) is a motivation only when it becomes part of our identity.

Our instincts motivate us for good or for evil. Radial insight (discussed in the next chapter), when properly developed, induces tangential insight to vector in the positive direction and induces a behavioral pull away from the irrational and towards good. Without effective positive modulation of tangential insight, our motivation becomes overwhelmed by our basic instincts. Radial insight induces spiritual behavior, the third mode of intentional behavior, be it perceptual, procedural, or psychical involves both an innate and a learned component.

Chapter 20—Sapient Stage

The sapience stage began with the appearance of the human mind. The mind introduced a different kind of reality than the other four creation events. The first four events are associated with the creation of new forms of matter: the universe, the earth, cells, and multicellular organisms. The main objective of this progression of increasing material complexity seems to have been the creation of material forms that maintained the life of its organizing cells while progressing toward higher complexity. Material complexity has reached its objective with the creation of the most complex material object in the universe, the human brain and with it the human mind.

Fossils of the first man-like creatures may to be 1 to 2 million years old; my interpretation of diverse sources suggests that the first anatomically modern human (AMH)—the *Homo Sapien Idaltu*—appeared about 160,000 years ago. The first true *Homo Sapien*, the Cro-Magnon, appeared about 50,000 years ago. Humans emerged late in the history of our planet and appear to be the endpoint of material evolution. As far as the human species is concerned, evolution has ended. I hold this to be true because the part of evolution that no longer applies is the "survival of the fittest." Technology, the result of the evolution of rational reality, has provided all humans with the ability to survive in all kinds of conditions and changes of environments. I contend that God directed the creation of the earth and life towards the eventual creation of a special animal, the human being. And here we are, and the sapient stage continues today as the sapient stasis period, otherwise known as history of humanity. The brain was the source of material complexity; but it was the creation of a mind that was the source of the quantum jump in consciousness. Although consciousness emerged from the brain, something else must be responsible for the mind. The common view states the obvious, "*the mind is an emergent property of the brain.*" Materialistic-leaning scientists use emergence as a curtain behind which they commence a search for the genii in the brain. Fortunately, enormous amount of useful knowledge about mental faculties and functions was generated behind that smoke screen to the benefit of humanity.

A. The Mind—Common View

We exist in an evolving ocean of matter and energy that has assumed a vast variety of forms and behaviors. The mind is capable of absorbing inputs (stimuli) and creating a vibrant panorama of sight, sound, smell, taste, and touching our own personal reality. However, being restricted to a common view, we can never define with specificity the lexicon of words used to describe the mental world, words such as: consciousness, thought, intellect, *cognition*, intentionality, and many others; they are too vague and interchangeable to provide distinct definitions. I deal with the complexity of the psychical world by attempting consistency with words that are often interchanged, words such as substance, faculty, function, property, and state. Just as "life" is described with a list of properties, neurologists also describe the mind by listing its properties, hence in Wikipedia we find: (i) "*the mind is a set of cognitive faculties including consciousness, perception, thought, judgement, language, and memory*", but, there is much more to the mind than that list of cognitive faculties. Another way the mind has been described: (ii) "*the faculty of entity, thoughts, and consciousness. It holds the power of imagination, recognition, and appreciation and is responsible for processing feelings and emotions, resulting in attitudes and actions.*" These two examples do not exhaust the words associated with the mind, words such as: concentration, reverie, decision-making, deviousness, volition, and others.

However, no matter the length of the list, the attempt to define the mind has only been descriptive, not explicative. Although there are descriptions, hypotheses, and definitions for the mind, none have been generally accepted. My favorite definition from the American linguist Steven Pinker: *"The mind is what the brain does; specifically, the brain processes information, and thinking is a kind of computation."* That concise statement captures the main elements in the study of the mind, namely, brain, information, thinking, and computation. But it leaves out, consciousness, the biggest mystery of the mind.

Before addressing that mystery, we must understand what we know about the faculties and functions of the mind. Central to the common view is the brain from which the mind emerges along with consciousness and thought. The brain and sentience are related through consciousness; sapience and mind are related through thought. Although the mind is one of the few mysteries with which all normal people are familiar; each one of us has one. Still, neuro-scientists—most of whom are normal—cannot resolve the mysteries of the mind, for example, we know that mind, thought, and consciousness emerge from the brain, but we do not know how. This is especially true of consciousness, the brain's most intriguing element. There is no way for another person, scientist, or philosopher to know what any other person's mind is like. We can only guess by studying on our own and assuming that the nature of our minds is like those of other normal people. Fortunately, the assumption of similarity is a good one because human life would be utter chaos if we did not agree that when someone says the apple is red, that red is what we, with normal eyesight, consistently see, even though the interior subjective experience may not be the same. Despite differences about how we think, what we see, what we hear, there are far more experiences upon which we agree. However, before diving into the mystery of the mind, let us first examine the faculties and functions of the brain, something that is well known.

B. The Brain

The brain is the most complex physical structure in the universe and if that is not incredible enough, the brain also gives rise to the mind, a mental faculty unique to humans. What we know about the mind is that the brain is necessary for it to exist. However, we suspect that the brain alone is not sufficient to create the mind, that there must be something else. Figure 12 provides a simplified model of the common view of the brain's central place in the faculty and functions of the mind. The brain is the sole source of *somatic, sentient,* and *sapient* systems.

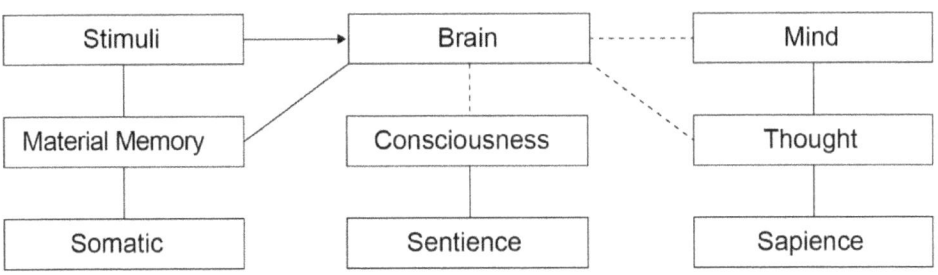

Figure 12. The Brain—Common View

Each system is composed of a faculty and a function. The brain and sentience are related through consciousness; mind and sapience are related through thought. The somatic system is different than the other two systems in that it is not an emergent property; it is directly connected. This organization implies two distinct roles for the brain: (i) as a faculty it is part of the nervous system and wholly understood to be completely material (nerves and neurons); and (ii) it is the source of sensory (non-material) functions. Before dealing with the two nonmaterial systems let us consider the material system, the somatic system.

C. Somatic System

The somatic system (i) is primarily composed of the nervous system that deals with the stimulus functions, (ii) consists of nerves that connect the brain and spinal cord to voluntary and skeletal muscles, (iii) operates solely through the nervous system, (iv) has been thoroughly studied and described, and is not considered as part of the mind-body problem (v) is the one system for which the brain is both necessary and sufficient.

The somatic system is the basic generation of physical behavior associated with skeletal muscles. It operates in material memory in both the common view and my view. Physical behavior is initiated by internal or external stimuli activating on the PNS in which afferent nerves send impulses to the CNS and then to the skeletal muscles. The somatic system is primarily a physical system and differs from the sentient and sapient psychical systems. The difference is described later when I discuss the sapient system, which resides in immaterial memory and makes mental decisions apart from physical stimuli. The somatic system operates solely through the nervous system and is wholly physical and not involved directly with the mind. The somatic system is the one system for which the brain is both necessary and sufficient.

D. Nervous System

The nervous system is composed of two parts: (i) the central nervous system (CNS), and (ii) the peripheral nervous system (PNS). Before diving into the mystery of the mind, let us first examine the faculties and functions that are, in the common view, called the *nervous system*. The nervous system is the body's control system; it provides three general functions: (i) it regulates vital functions, (ii) it supports homeostasis to maintain a constant internal environment, and (iii) it is essential for human sapience. The nervous system consists of two major faculties: (i) the central nervous system (CNS) composed of the nerves in the brain and spine; and (ii) the peripheral nervous system (PNS), composed of the cranial nerves and those that exist outside of the brain and spine. The CNS is associated with the *motor division* as an output function, and PNS is associated mainly with the *sensory division* as an input function.

1. Central Nervous System

The central nervous system consists of the spine and the brain. The brain makes up the largest portion of the CNS. It is often the main structure referred to when speaking of the nervous system in general. The CNS is referred to as "central" because the brain integrates the received information and coordinates and influences the activity of all parts of the body. In vertebrates, the CNS also includes cranial nerves that bring information to the CNS to and from the face, as well as to certain muscles (such as the trapezius muscle, which is innervated by accessory nerves as well as certain cervical spinal nerves. Apart from the spinal cord, there are also peripheral nerves of the PNS that synapse through intermediaries or ganglia directly on the CNS. These nerves exist in the head and neck region and are called cranial nerves. Cranial nerves bring information to the CNS to and from the face, as well as to certain muscles (such as the trapezius muscle, which is innervated by accessory nerves as well as certain cervical spinal nerves.

The sensory division contains *afferent* neurons that react with different types of stimuli and send action potential to the brain via the spine. Sensory neurons are nerve cells that send information from sensory organs to the central nervous system (CNS) about changes in the body's internal and external environments. It consists of two systems: the somatic and the autonomic.

The *autonomic nervous system* generates involuntary behavior, which can occur without our being awake. This system operates the internal organs. The homeostasis mechanism, the self-regulating mechanism that maintains an optimal internal environment by controlling internal body temperature, tissue fluid, sodium, glucose, and composition of the blood and other necessary life-sustaining factors also generates internal stimuli. Innately constructed neuron subsystems of the central nervous system moderate homeostasis. Neurologists divide the autonomic system into three components: (i) the sympathetic, (ii) the parasympathetic, and (iii) the enteric divisions. The *sympathetic* division prepares the body's systems whenever a stimulus requires dramatic behavior (fight or flight). The parasympathetic system moderates the systems when the body is at rest. The *enteric* division governs the functions of the gastrointestinal tract. The sympathetic and parasympathetic systems involve subjective experiences that are not muscle related, pain for example. Because subjective experiences occur in the mind, and are unobservable, science knows the mind only regarding its relationship to the brain.

The *motor division* generates voluntary behavior through the skeletal muscles. External stimuli excite sense receptors that send sensory impulses along the afferent nerves of the PNS to the brain part of the CNS where analysis of the information decides to send impulses through efferent nerves to the appropriate muscles. Repetition of such voluntary behavior, either innate or learned, creates permanent neural maps. Neurologists refer to the operation of the somatic system as *motor* or *muscle memory.*

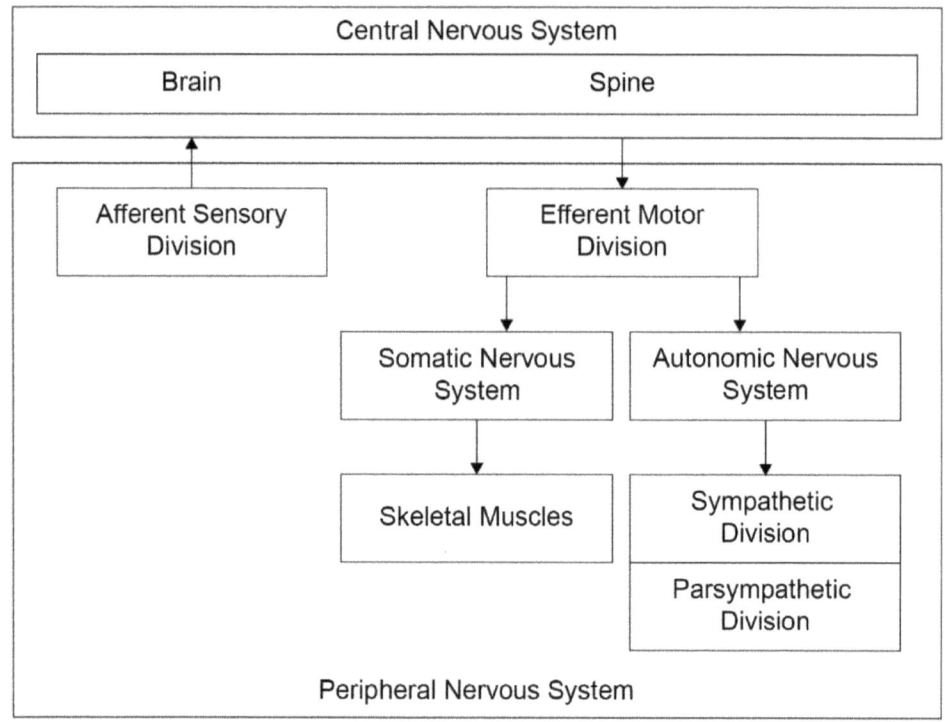

Figure 13. The Nervous System

E. Memory

In addition to describing the main faculties of the brain, science has also had much success in describing at least one function of the mind, memory. In the common view, neural maps are both necessary and sufficient to explain all an animal's—including a human's—subjective experiences. Science claims that subjective experiences such as qualia and feelings are emergent properties of the neural map in the brain. Neurologists have isolated the areas of the brain where many subjective experiences occur, and this is compelling evidence that the brain is necessary but there is no evidence that the brain is sufficient. Science has also described the functions of memory. A common approach is to identify the main types: sensory, working, or short term, and long-term memory. Some memory theorists conflate working and short-term memory, while others use the terms to differentiate or eliminate one or the other. I will treat them as separate functions.

Sensory memory is the first step in any description of memory. It refers to the brain's capability of briefly holding stimuli from the sense receptors. Sensory memory is associated with four types each associated with a different sense, they are: iconic memory (visual stimuli); echoic memory (sound stimuli); haptic memory (touch stimuli); and proprioceptive memory (proprioception). The storage of the stimuli is less than a second except for the echoic type where it takes 2-3 seconds to process sound after which the information transfers to the working or short-term memory.

Working (short-term) memory is the retention of an input from the senses for less than a second to allow the brain to select information that, according to the common view, the *brain* decides is worth transferring to the long-term memory. The typical amount of information held in short term memory is 7–9 items; retention is a few seconds duration without further *memorization,* i.e., (i) *rehearsal* (constant repeating) or *chunking,* (ii) (organizing the information in groups). Although the working memory is the subjective experience that we call *attention to,* the subjective experiences such as interest, curiosity, or orderliness make the decision to transfer to the long-term memory.

Long-Term memory stores information from the short-term memory for an extended duration, the amount of which is retained forms the *intellect.* There are two parts: (i) *implicit* (or procedural) and (ii) *explicit* (or declarative); declarative memory consists of *episodic* or *semantic* memory.

Implicit memory unconsciously and spontaneously recalls stored sequential information (fluency). For example, recalling the words of a poem or a song, the multiplication table, or riding a bike. Science knows the parts of the brain, namely the basal *ganglia* that encode information but are not sure where it is stored.

Explicit memory has two forms: e*pisodic* memory stores percepts formed from events and images; semantic memory stores concepts, facts, and meanings. Explicit memories are consciously available. The *hippocampus* and the *rhinal cortex* encode percepts and concepts, and are stored elsewhere? The last statement, "they are stored elsewhere" is a clue to the shortcoming of the common view of the mind; for where is the "elsewhere"? The current common view is monistic, a view that restricts 'elsewhere" to the neurons somewhere in the brain (association areas?); but if that is true neurologists would have already found their

---

Note 37: *association areas* are the parts of the cerebral cortex that are not part of the primary regions. They are involved with subjective experiences and support abstract thinking and location in the brain. The description of the material memory primarily addresses language.

---

physical behavior that has been thoroughly studied and explained so here we have arrived at the so-called mind-body problem, the inability to deal with subjective experiences such as consciousness and qualia. The view of mind restricted to the material memory works well for describing physical behavior but is trapped in a maelstrom of attempts at explanations that keep busy the experts that deal with the mind searching for answers to the subjective experiences.

1. Sentience

Sentience is the mind's experience of sensations such as sight, sound, smell, taste, touch, feeling, and emotion. Explaining and even describing sentience is impossible, it can only be experienced. Science is stuck with the abstraction "it emerges from the brain." I will go a step further and use the stuff the ancient Greeks called "nous" to explain sentience. The sentient system consists of brain faculty and the consciousness function

2. Sapience

The sapient system is composed of the mind faculty and thought is its function. Whereas sentience acts in response to events external to the mind, the sapient faculty is a part of the mind, the nous, that is self-stimulating. Sapience is usually meant as a synonym for wisdom. I define it as an output faculty in which thought can make decisions to act; it is a sapient faculty. It has three functions: thinking, thought, and cognition. We learn with thinking and thought to create conceptual intellect, and we use cognition to create the perceptual intellect. To understand these mental faculties better we must turn to immaterial memory. The mind is a dualistic faculty. Hence, it is much easier sorting out the structure and activities of the mind if we let the material memory deal with the physical elements and allow the presence of an immaterial memory to deal with the psychical. Material memory—the brain, spine, and nervous systems—has been thoroughly studied and understood; the same cannot be said for immaterial memory. In fact, immaterial memory does not exist in the common view. The hylomorphic dualism, I argue exists, and provides a way to describe an immaterial memory, the source of the physical elements such as consciousness, which can't be explained by materialistic science

The common view of the mind in which the mental faculties all derive from the brain leaves unanswered questions dealing with elements of subjective reality such as: consciousness, emergence, sentience, sapience, free will, and volition. They will remain unanswered as long as the scientific paradigm is a prerequisite for truth and the hylomorphic dualism is disallowed. Despite the obvious dualistic nature of the mind, the main problem for those that study the mind is that only the brain can be considered. Based on my premise that God exists, I suffer no such handicap, and I take advantage of a dualistic approach that the existence of a psychical substance permits. I have two remarkable tools to work with: (i) nous, the psychical substance, and (ii) an immaterial memory, a psychical faculty. I use them to unravel the bewildering common view of the faculties and functions of the mind. Hence, if we approach the mind-body problem armed with the existence of a psychical substance, there is no mind-body problem.

F. The Mind—My View

The mind is a hylomorphic faculty consisting of both the material and the immaterial based on discrete and continuous space. The material part of the mind resides in the brain, specifically, the Broca, Werneke's, and other areas that form the language faculty. The immaterial memory resides in the nous. One of the advantages of the inclusion of immaterial memory is that it allows a more specific understanding of the psychical elements that form subjective reality. Nous provides the duality that allows us to define the mind as *the interaction of the human language faculty in the brain with the nous, a psychical component in which the brain is immersed.* Since the language faculty is composed of discrete elements and nous is a continuous substance, the mind is a digital-analog structure. The digital-analog structure allows us to find plausible answers to science's unanswered questions such as the mind-body problem and the nature of qualia. Given an explicative definition of the mind, we can better understand how the mind works.

The mind as shown in figure 14 is hylomorphic in that it is composed of the language faculty in the brain combined with an immaterial memory. The mind formed in the *nous*. Nous, the main difference between the common view and my view, allows me to describe an immaterial memory that acts as the faculty in which subjective experiences are processed. The common view of the mind—in which the mental faculties all derive from the brain— leaves unanswered questions dealing with elements of subjective reality such as: consciousness, emergence, sentience, sapience, intentionality, free will, and volition. They will remain unanswered as long as the scientific paradigm is a prerequisite for truth and the hylomorphic dualism is disallowed. Despite the obvious dualistic nature of the mind, the main problem for those that study the mind is that only the brain can be considered. Based on my premise that God exists, I suffer no such handicap, and I take advantage of a dualistic approach that the existence of a psychical substance permits. I have two remarkable tools to work with: (i) nous, the psychical substance, and (ii) an immaterial memory, a psychical faculty. I use them to unravel the bewildering common view of the faculties and functions of the mind. Hence, if we approach the mind-body problem armed with the existence of a psychical substance, there is no mind-body problem.

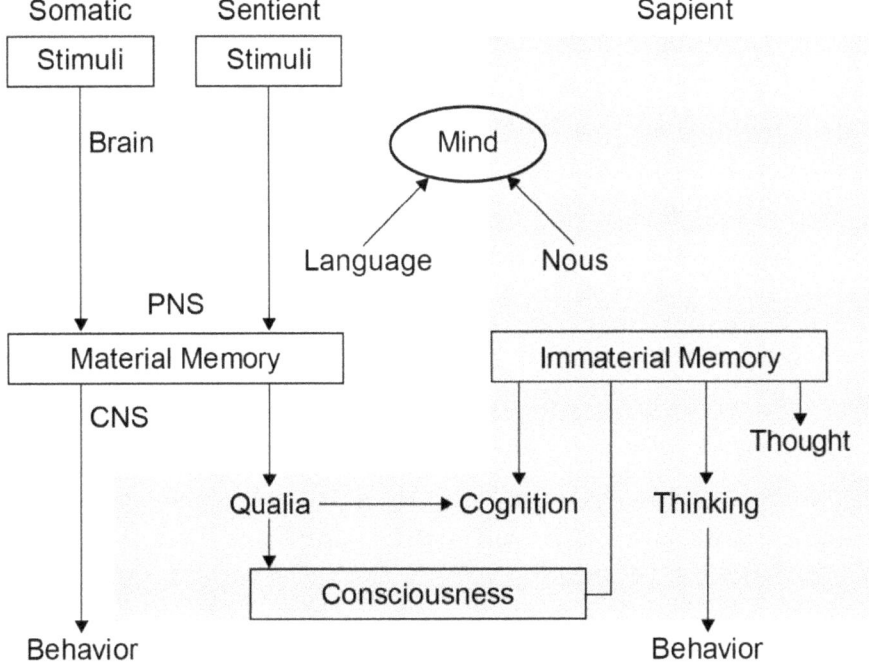

Figure 14. Human Mind—My View

Note 38: Animals possess nous but not language; they are sentient but, by my definition, do not have human minds. Because animals are sentient and experience qualia, they have what might be called pseudo minds.

Chapter 21—Mind-Body Problems

In the common view, the phrase "*the mind-body problem*" refers to the relationships between mental states and neural states, between mind and brain, between immaterial and material. There have been many proposals offered as solutions, but none has ever been recognized as an acceptable theory. The result seems to be that the mind-body problem has resolved simply into neuroscience's inability to find an acceptable explanation of consciousness. The general practice has been to make consciousness the primary subject for investigation with the confidence that explaining consciousness will explain all the faculties and function of the mind especially the process of how mental states emerge from the brain. The mystery of consciousness creates substantial interest among neuroscientists, philosophers, biologists, psychologists, and amateurs like me. No matter how divergent the results have been, the one common observation held by all is that the mind and consciousness emerge from the brain. Emergence is neuroscience's favorite abstraction. However, the mind is far too complex a faculty to dismiss with a simple hand-wave that contends it emerges from the brain, so the search continues, and its objective is a plausible theory of consciousness. There are two main schools of thought into which the approach is divided: monism and dualism. There are as many theories of *monism* as there are people of interest thinking about the idea. A partial list would find the following monism types of substance, existence, ontological, partial, priority, genus, and property monism. We can sort the extensive list into three more general categories: idealist, neutral, material monism. (i) idealistic monism means that only the psychical exists; (ii) neutral monism means that only one substance reduces to both the physical and the mind; and physical monism (or materialism) means only matter exists from which the mind emerges. All the versions of monism have one thing in common; all reality derives from a single thing and there is no need for a psychical substance. Functionalism, physicalism, identity theory, behaviorism, neural group selection, and others are hypotheses of the mind that all have in common the denial of dualistic memories, material and immaterial.

Dualism also appears in several versions. Mind-body dualism contends that the mind involves both a material and an immaterial source. The proposal presented by Descartes took the view that there are two distinct, independent, and irreducible substances, one forming the material, the other forming the immaterial. Cartesian dualism died for the lack of an explanation of causal interaction, the problem of explaining how the physical and psychical impact one another. Cartesian was replaced with several other dualism hypotheses such as: substance, property, predicate, interactionism, and non-reductive dualism. A fashionable explanation called the computational theory of the mind describes the nervous system as an information processing system and (i) the mind is the flow of information through the nervous system and (ii) consciousness is the "experienced information flow." This approach presumes to separate the material (nervous system) from the immaterial information and thus introduce duality without the need for a psychical substance. That just adds another layer of abstraction (information) to the problem and treats consciousness as a "substance." There are many other views, none of which has become a universal solution to the mystery of consciousness.

Because the present scientific paradigm restricts its view of the mind to a materialistic solution, those in the field are confronted with unsolvable problems dealing with the fact that consciousness, the most intimate experience of the human mind, is immaterial. Despite the enormous amount of work that went into solving the mind-body problem, the effort to describe subjective experience as a physical, material nature has resulted in a failure to find an accepted solution.

A. Emergence

This is an abstraction that does the job of describing observations but does not explain how qualia interacts with the neurons. I am certain that a materialist would not accept the existence of the nous, let alone the claim that it is the source of subjective experiences; on the other hand, a materialist accepts the variety of physical fields such as: electromagnetic, gravitational, Higgs, and dark energy that arise in space as the result of specific material conditions (flowing electrons for example). Science has no trouble assigning non-visual properties such as dark matter to space but only when observed under specific conditions. If the motion of a material electron creating an immaterial, magnetic field is a scientific fact, why cannot the motion of charged material ions at neural synapses in the brain create an immaterial subjective experience of color? It, of course, can. Nevertheless, there can be no denying that amazing things go on inside our mind, the most amazing being expressed in the question, how does the material brain impact the immaterial mind? I use hylomorphic dualism to explain emergence and in doing so offer a solution to the consciousness problem. Before dealing with consciousness, I first deal with two other mental mysteries that are seldom mentioned; they are transmission and reflection.

B. Afferent Body-Mind Problem

The hard problem of consciousness, in the common view, is a "body-mind" interaction in that information flows from the bodily sensory nerves to the mind . However, given the presence of the nous, and the s-configuration that defines the morphic s-frame (body), the body-mind interaction in which material affects immaterial in an afferent event is, as pointed out earlier, not any more mysterious than the phenomena in which matter in the form of electrons moving in a copper wire create a magnetic field in space surrounding the wire. In the common view, qualia are emergent properties of the brain. That the brain is necessary, and qualia emerge from it, is certainly true. However, that which emerges must have somewhere in which to emerge. The common view has not produced an acceptable answer to that question. The existence of nous and immaterial memory does allow a plausible explanation. In this section, I address the body-mind afferent mechanism, how the neurons activate qualia.

C. Mind-Body Problems—My View

To find a plausible explanation for consciousness we must first define our mental terms with specificity, but this is only possible when we accept the existence of an immaterial memory. Only then can we avoid conflating consciousness with terms such as sapience, sentience, awareness, and thought. In addition to being more specific with other mental terms, For example, the phrase "mind-body problem," in the common view, lumps both the hard and the weak problem under this one phrase. In the common view, the hard problem of consciousness is finding an explanation for the mental experience of qualia, our sensations, and feelings. This hard problem of consciousness is an afferent event in which external information stimulates the peripheral nervous system (PNS) that sends impulses to the brain that activates the qualia contained in the nous in the immaterial memory. The problems associated with this afferent input event would be understood more specifically if it were referred to as the "body-mind" event and use the mind-body phrase to describe the efferent output event I call the *sapient function*.

My view is based on a different structure of the mind. Whereas the common view is generated with a single faculty (the brain), my view of the mind-body problem is simplified by: (i) the psychical substance infinite nothingness and its modalities pneuma and nous; (ii) by the formulation of a hylomorphic dualism; and (iii) the existence of an immaterial memory.

The existence of an immaterial memory allows simplification of the consciousness mind-body problem, a problem I refer to as an afferent problem. An immaterial memory implies a dual function of the mind: (i) an afferent function in which information flows inward from the material memory to the immaterial memory (mind) where consciousness is activated and (ii) an efferent function in which information flows from the mind to the material memory where thought and intentionality are activated. The existence of an immaterial memory allows possible solutions for three other mental problems. I refer to them as mind-body problems; they are: (i) the *transmission problem*; (ii) the *binding problem*; and (iii) the mind-body efferent problem.

Usually, consciousness is defined in two parts such as: Ned Block's P and A consciousness and David Chalmer's hard problem and weak problem of consciousness. Block's P-consciousness and Chalmers' hard problem of consciousness are focused on the qualia problem, and Block's A and Chalmer's weak problem serve to address all the rest of the mental functions such as: (i) information, (ii) representations, (iii) intentionality, (iv) thought, (v) speech, (vi) reason, and (vii) control of behavior; by discovering the neural coordinates in the brain.

Contrary to the common view of the afferent body-mind situation as the "hard problem of consciousness," I consider it to be just one of four major problems not solved in the common view but can be explained with the explicative ("my view"). In addition to the afferent body-mind problem, I also address the weak problem because the abstraction "emergence" is just as plausible as "magnetism" or any other of the abstractions currently in use for describing observations. Hence emergence suffices as a plausible explanation.

The common view does not apply a catch-all abstraction like emergence to describe the efferent mind-body faculty such as thought or volition, hence there is no clear description for emergence. Hence, the mind-body efferent problem (thought) is really the "hard problem of consciousness" because there is no convenient abstraction solution that describes mind-body experiences. The mind-body efferent problem can be stated as: how does the mind, an immaterial faculty, cause material neurons to initiate behavior?

I only know how my mind works, and the premise that God exists allowed me to propose a hylomorphic mind based on the presence of nous in which to posit an immaterial memory that contains the qualia. Visual illusions and, for example, finding an answer to the binding problem is easier to explain when dealing with a reality evolving incrementally 10^{+43} times per second in an infinitude of layers of psychical substance. Incrementation and infinitude of layers suggest a greater latitude for discovering the nature, location, and amplitude of every quale in each percept stored in the immaterial memory. The multiplicity of increments allows sequencing of sensations so that the various elements in a percept arrive in the mind sequentially, remembering in my view the increments of the cosmic s-frame are arriving every 10^{-43} seconds.

162

1. Reflection

When looking in a mirror it dawned on me that: (i) I could not see my face but knew where it was and (ii) its image was in two places. I asked myself could the image of my face be in two places at the same time. Light reflected from my face and travelled to the mirror where it was reflected to where its image was created in my mind. Light is reflected perfectly only from smooth, and plane surfaces like we find in cosmetic mirrors. The reflected image is flat (2-dimensional), as is the transmitted configuration. When the transmitted image is processed in the mind, it is perceived as 3-dimensional and by extromission, is retransmitted to the mirror instantaneously.

What I am demonstrating here is the power of hylomorphic dualism for explaining otherwise inexplicable attempts that were restricted to neurons. Since the pneuma is simply infinite nothingness that is conterminous with the universe there is communication between the infinite nothingness and the pneuma. The same is true of pneuma and nous in immaterial memory. Hence, communication between the mind and the rest of the universe and beyond. Communication between modalities of the infinite nothingness is a form of "transcendence," a connection between the nous in metazoa and the noumena in the pneuma.

2. Transmission

I know that the object (an apple) across the room is nothing, but a configuration of s-points formed as quarks and electrons that reflect light that is nothing but photons. And what I see across the room when the reflected photons that enter my eye and are perceived in my mind is a round shiny red apple. *The transmission problem is explaining how the image of a red apple that is perceived in my mind*, appears across the room on a configuration of quarks and electrons. The ancient Greeks addressed this problem. Democritus' theory of intromission assumed that vision resulted from small particles emanating from the object to the eye. Plato went to other way, his theory of extromission assumed some visual power emerged from the eye and transposed its image to the object. Both are partially right because in my view, the image travels in both directions.

As light is reflected from the apple, because of the incrementation of the cosmic s-frame, the *configured light* reaches the eye as sequence of photons where they are focused on the retina at the back of the eye. Here the photons are transduced into electrical pulses and sent to the visual cortex through optic nerves and an image appears in the mind. The mind not only "sees" the red apple but sees it across the room. Obviously, what we see in our mind is transmitted back to the object through the pneuma instantaneously. The speed of intromission in discrete space of the reflected configuration is the speed of light, but the speed of extromission in the continuous pneuma is infinite. Hence, the image formed in the mind appears on the apple instantly, and we think (*intend*) the apple across the room is red

> Note 39: The speed of light in discrete space equals the Planck length ($\sim 10^{-35}$ meters) divided by the Planck era ($\sim 10^{-43}$ seconds). However, if the light travelled in continuous space (like the pneuma) the era = o and the speed of information is infinite.

In thinking about this transmission problem, I realized there are two other visual events that must be dealt with; (i) reflection, the image reflected from a mirror, and (ii) projection, images projected through film appearing on a screen. Hence: for the transmission, reflection, and projection events, the parts in all three visual elements are the same; they are: the object, light, the eye, and the mind.

3. Nous

Nous (Greek for intellect) is a psychical substance, hence disallowed by science because it leads to the existence of God and that is not part of the modern scientific paradigm. However, the "God-exists" premise allows the existence of nous; that allows the existence of an immaterial memory, and what that allows is a means to explain subjective, rational, and transcendental reality.

D. Immaterial Memory

The immaterial memory is formed in the nous part of the mind. It allows an explanation of emergence, the common view of consciousness. The brain is necessary because the mind depends on syntax, the words, and their arrangement, which are stored as a language faculty as part of the material memory. The meanings generated by language—semantics—reside in the immaterial memory. The brain is necessary to engage the material memory but is not sufficient; the nous is necessary to engage the immaterial memory but not sufficient; the mind requires the presence of both to be necessary and sufficient.

Given the experience of consciousness, it should be apparent that consciousness is not material, that it derives from some form of immaterial substance. In other words, to explain consciousness, materialism will not work because the brain is not sufficient, and because a brain is necessary, idealism will not work; so, we are left with dualism.

I begin with what I believe about consciousness. My first belief is that consciousness is not fundamental. In the common view, the word consciousness is presented as fundamental and is often presented as leading to panpsychism. I contend that nous is the fundamental element of subjective and rational reality; and pneuma is a fundamental element of objective reality. Although it can be understood both as a state and as a property, consciousness is not a fundamental substance.

E. Efferent Mind-Body Problem

Immaterial memory provides a way to find answers to the questions that science does not have. The unanswered questions are associated with the nature of behavior, thought, mind, and soul. To address these questions, we must unravel the nature of the human mind. I present an argument based on a hylomorphic structure of reality that utilizes the psychical substance called *nous* to explain our subjective experiences. The immaterial memory resides in the nous that combines with the language faculty to form the mind.

Neurobiology deals with both a mind-body and a body-mind problem and both are "hard problems." Because some of the neural channels of spontaneous mind-body behavior run from sense receptors to the activation of muscles is called the weak problem of consciousness to hide the fact that much of the mind-body behavior originates in the mind not in the stimulus. But what and where does intentionality arise—in neurons? The only way to explain intentional mind-body behavior is to postulate an immaterial memory in which a sapient, decision-making faculty exists.

An *efferent event* begins when the material memory is activated, and neurons send impulses through the efferent nerves to carry impulses to muscles. There are two paths: one originating in the sensory system and another originating in the nous: for impulses that originate in the sense receptors, travel directly through the PNS and CNS; and arrive at the muscles. There are neural maps for every motion of a muscle or any combination of muscles in the human body. Learning and repeating bodily muscular movements create specific neural maps called motor memory. The journey of impulses from a synapse in a sensory neuron is well understood. Events that originate in the nous raise questions as to how the immaterial mind activates a neural map.

The afferent body-mind event in which neural (material) maps activate immaterial qualia fields is analogous to the creation of a magnetic field by electrons flowing in a copper wire. Similarly, the efferent mind-body event in which immaterial mental states activate material neural maps is analogous to the physical event in which a wire moving through a magnetic field causes electron flow in a wire. Hence the sapient faculty can initiate behavior of matter without external stimuli.

The common view of the afferent event being the "hard problem of consciousness" can, at least, be hidden under the emergence abstraction. We can always declare that redness is an emergent property of the brain. But how do we attribute something that is immaterial to the power of emergence? The materialist is left with the argument that if there is no immaterial faculty to make decisions, neurons must.

Subjective experiences are inexplicable because of science's insistence that only matter, and energy produce the mind, and the material brain suffices to produce emergent properties, and there is no need to create an immaterial substance to explain these unanswered questions. On the other hand, if science can invent dark matter to produce a plausible explanation of an observation, and then argue that the explanation verifies the existence of dark matter, why not invent dark matter to explain the mysteries of the mind? Let us see how that works.

In my view, material memory deals solely with *the somatic functions*—that are initiated by the sensory system. Sentience and sapience can be explained only with the existence of the nous that forms an immaterial memory in the mind. I am now equipped with some new tools, namely sentience, sapience, material memory, immaterial memory, nous a psychical substance, and an explanation of the material-immaterial interface, with which to address the mystery of consciousness. Another subject that can be addressed with the existence of an immaterial memory is the abstraction often used in the common view, the word emergence.

F. Emergence—My View

Since the immaterial memory and the brain are coterminous, the presence of the nous explains emergence. There is an innate connection between the language faculty in the brain and the immaterial memory in the nous. The brain is necessary because the mind depends on syntax, the words, and their arrangement, stored as a language faculty as part of the material memory. The meanings generated by language—semantics—reside in the immaterial memory. Hence the mind "emerges" from the brain because the semantics that are essential to the mind depend on the syntax that that is created by the language faculty in the brain.

With an immaterial memory, we can describe the mechanism in which the sensations entering the brain are transformed into electrical impulses that cause synapses to activate a specific signifier field just as electrons moving in a wire create a magnetic field. Each quale is associated with a specific depth of the nous -a qualia field - that is contiguous with the neural map in which several specific synapses are located. Each q-map is associated with a specific quale, for example, redness is experienced when a specific q-map is activated by a specific signifier field or by a single neuron. The color and brightness are determined by the nature of the q-map. and the number of activated synapses.

A quale's identity is determined by the number of synapses in a particular quale; the number of synapses in a quale determines the density of nous for each quale; the number of synapses in each q-map (the area in the brain, in which they are determined by the amount: each quale is identified by density and location of nous; each q-map is coterminous with the activating neural map; each density of nous is determined by number of synapses in neural map. I am not claiming that the above is the mechanism that solves the body-mind problem; I am simply presenting this as a plausible argument derived by assuming the existence of God as a premise, and as an explanation what happens in a physical sense. The excitation of the qualia field by an active neural field is called: an afferent, (ii) a body-mind, or (iii) a conceptual event. The mind is also involved in two other events: (i) the efferent mind-body, (iii) or a perceptual event.

One should recognize that what I am presenting as is the mechanism David Chalmers refers to as the "hard problem of consciousness" and Ned Block calls p-consciousness. The general common view uses the abstraction of emergence to explain the experience of qualia, but no general agreement exists regarding the efferent mind-body associated with sapience. This problem called the sapient faculty, is shown on the sapient side of the immaterial memory chart.

Psychogenesis began a new stasis period in which the human mind was the dominant element that allowed sapient behavior. Human minds are the elements of a sapient stage just as physical elements defined as the cosmological stage; molecules the geological stage; cells the biological stage; and animals the sentient stage. Before psychogenesis, consciousness was associated with material complexity, after psychogenesis, complexity depended on human consciousness

G. Sapient Stasis Period

The sapient stage began with the creation of the first human about 50-thousands years ago. In comparison to the first four stages, the sapient stage is but a blink of time. Psychogenesis, the actualization of the mind, introduced the sapient stasis period from which rational reality emerged. The human being, more specifically the human mind, is the basic element that created culture. The presence of the mind added enormously to consciousness in an incremental way and to material complexity in a more gradual way. However, the sapient stasis period, more simply, is human history. Just as previous stasis periods began with and evolved towards a creation event, so too the sapient stasis period was progressing toward a similar event. However, the nature of such an event is unlike the first four events that were characterized as quantum jumps in material complexity; this event was characterized by a quantum jump in consciousness. Consequently, such an event cannot be imagined, we can only surmise that such an event existed because there is a progression that necessarily leads to an objective that in a God driven world generates hope.

To discern that objective, we must analyze the nature of progress for which a plausible metric is goodness because hope is always bound to good. Unlike the previous stasis periods during which material complexity evolved, in the sapient stasis period culture drove increasing consciousness.

Consciousness separated from material complexity when observable biological evolution ended, and the evolution of consciousness continued because of the evolution of the human mind. Since then, the only increase in material complexity has been associated with the unknown and dubious increase in the number of species and with the material complexity created by humans through technology. Consciousness, on the other hand, which had tracked material complexity since sentience appeared, detached at a rate identified as a creation event after which material complexity increased gradually while complexity of consciousness followed an increasingly higher rate. To understand consciousness, as an element of increasing good, we must examine the cause of consciousness separating from material complexity an event that is obviously associated with the human mind. This led me to examine human behavior. To start I recognize that we humans behave at three levels: animal, rational, and spiritual. The three levels are all organized around the schemes shown in Figure 15

H. Noosgenesis

Noosgenesis, like the previous creation event (Psychogenesis) did not include a dramatic increase in material complexity. The seminal event of this creation event was the humble birth of a child in a lowly manger in an insignificant corner of the world. The baby was named Jesus of Nazareth, God made man. Like prior creation events, the seminal event is followed by a relatively short period of rapid growth before an objective is reached. For example, the big bang was a seminal event that took 100 million years before the first star began to shine and an objective (the universe) appeared. It was at that point that we can say that the universe was created. The Noosgenesis creation event (seminal to objective) began with the birth of Jesus and reached its objective 380 years later when Constantine made Christianity the religion of the Roman empire. The seminal event was followed by a relatively brief period of rapid growth before the objective was reached and has been progressing over 2000 years towards the Omega Point. That implies that we are now living in a spiritual stasis period heading for the Omega point. My view of the Omega point is not as esoteric as Teilhard or Tipler's. Because of the idea of sequential lives in parallel worlds our bodies are incrementing and will be at the omega point when goodness reaches a peak in which evil is extinguished.

There are three forms of evil: natural, social, and personal. It is difficult to envision the attenuation of natural evil or the universal decline of personal evil. But God's teleological plan is to eliminate the possibility of evil and that will include all evil meaning all three-types. The Omega Point, the objective of the noosphere must be the creation event in which social evil has virtually been eliminated. Then during the spiritual stasis stage, with the cooperation of nations, natural evil will be eliminated in accordance with our submission to the Cosmic Christ. Finally, we reach the pleroma when all personal evil is distinguished, and all souls have been justified and centered in the Cosmic Christ. That said, I can now imagine the Omega point will be a perfect world immersed in perfect justice. This view portends the establishment of permanent world peace setting in which humans can work together to eliminate personal evil. The Omega point will be: (i) total world peace, (ii) complete social justice, (iii) perfect environment and (iv) all evil eliminated.

Chapter 22—Spiritual Behavior

In Part 4, I use the results found in the first three parts to find plausible arguments to deal with the questions that materialists, skeptics, and cynics—the secular humanists—confront theists. We theists normally address non-believer's questions by invoking faith. But faith is a weak weapon with which to confront: (i) the materialist that invokes scientism; (ii) the skeptic that invokes the absence of proof; and (iii) the cynic that evokes the existence of evil? To enhance faith and to support the belief of the heart with the belief of the mind, I utilize the other two theological virtues—hope and charity— to support faith.

A. Spiritual Corpus—The Soul

The mind is the psychical faculty that transforms the human animal into a rational person. The soul is the psychical faculty that transforms the rational human into a spiritual person. It is a person's behavior that forms the soul. In fact, the soul is the ongoing transformation of personhood, the temporal element of the soul.

In abiogenesis, matter captures and isolates a particle of the pneuma. The encapsulated pneuma, an immaterial substance, within a cell is called bios. Bios provides the means for the holonomic mechanism to operate at the cellular level and to animate the cell. The idea that crossing a threshold of complexity is important because the bios effectively controls the cell and brings it to life. The relationship between consciousness and material complexity foretells the relationship between the spirit and soul. Just as the soul enlivens the multicellular animal; we might think of bios as a pseudo soul that forms the cellular s-frame that in turn forms the spiritual substance (pneuma) to create the pseudo soul, or as Aristotle called it, the vegetative soul.

For each person, the path of righteousness is an evolution of the pseudo soul through the rational soul to the spiritual soul. The three modalities of the soul, animal, rational and spiritual evolve for humanity along both the path of actuality and for individuals along the path of righteousness. Along with the POA, the vegetative soul appeared during abiogenesis in the form of bios when it introduced life. The rational soul appeared during somagenesis in the form of nous when it introduced sentient life. The spiritual soul appeared during psychogenesis in the form of the mind when it introduced sapient life. and actual grace works through the mind to induce charity.

The spatial sequence determines our centrality and centrality determines the power of selfhood in motivating our day-to-day behavioral decisions. Since the spatial sequence is associated with the sense of internal proprioception, i.e., with our body, and the body induces the needs of identity, security, and stimulation, selfhood is a manifestation of the basic instincts. Basic instincts are the motivation of animal behavior; it is a separate element not shown in the immaterial memory chart. The motivation element shown as a component of volition refers only to rational and spiritual behavior discussed below. Nevertheless, basic instincts are the most powerful level of motivation. It is the negative side of the soul. The temporal sequence determines personhood and modulates our basic instincts. Its effectiveness depends on cognition that includes rational and spiritual motivation, volition, and free will. Whereas selfhood is innate, the development of personhood appears as constant thoughts flood the mind about every waking moment of every day. What should I wear? What should I eat? What should I do? Should I read or watch TV? And, on and on and on. The answers to such mundane indecisive choices, which have inconsequential outcomes, illustrate the existence of a duality in the mind that is also present when the questions have consequential outcomes. It is this contest that determines the relative weight of personhood and selfhood, the integrity of the soul. It is most beneficial to the integrity of the soul when it is the rational motivation that makes the decisions even when the rational decision is to allow selfhood to motivate the act.

The soul is a unitary faculty consisting of personhood and selfhood and there are two ways to imagine the soul; (i) as centrality, the separation of the personal psychical center from the cosmological center, or (ii) as the ratio of personhood to selfhood. Imagine a pie chart that begins with 100% selfhood, and then sometime later a different colored segment (personhood) appears and grows, we can say the integrity of soul growing towards justification. The soul is the relative difference between personhood and selfhood. What I am alluding to here is the idea of *hypostasis*, two natures in a single body. Although made in the image of God, human hypostasis does not include a Divine Nature as does Jesus. The dogma concerning the hypostatic union states: *The Divine and human natures are united hypostatically in Christ, that is, joined to each other in one Person.*

B. Spiritual Stimuli—Grace

There are different forms of grace, for example, you might find grace described as: sanctifying, actual, efficacious, efficient, sufficient, justifying, gratuitous, sacramental, and a couple of others of which I am not aware. The definition of each is fair game for any theologian that cares to address the subject and consequently there is overlapping of meanings. However, to make sense of the subject and make it forensically useful, I simplify by recognizing just three general kinds of grace: habitual, sanctifying, and actual. Usually habitual and sanctifying are interchangeable, but I separate them based on the approach described in the opening paragraph of this chapter that describes the three ways that we seek God: upward, inward, and outward. Habitual grace is present at birth; sanctifying grace is present in the worship of God; actual grace is present with justification.

We recognized the type of grace by the effect it induces. Habitual grace induces awe; sanctifying grace induces peace, and actual grace induces joy. God rewards us when we seek goodness. Grace is effective only when accepted and utilized. We recognize grace through recognition and/or through inspiration. We find it in subtle portions by searching for and recognizing it in our own behavior. The assent to grace is associated with free will; if God infused grace into our souls, and we accept it, we could not make a choice other than for good, and we then could not do our job of doing good and eliminating the possibility of evil

When God created free will while endowing humans with the motivation of our animal needs, life became an eternal struggle in choosing between good and evil. God also armed us with the grace to motivate the free will to control such enticements and to decrease evil and increase good in the world.

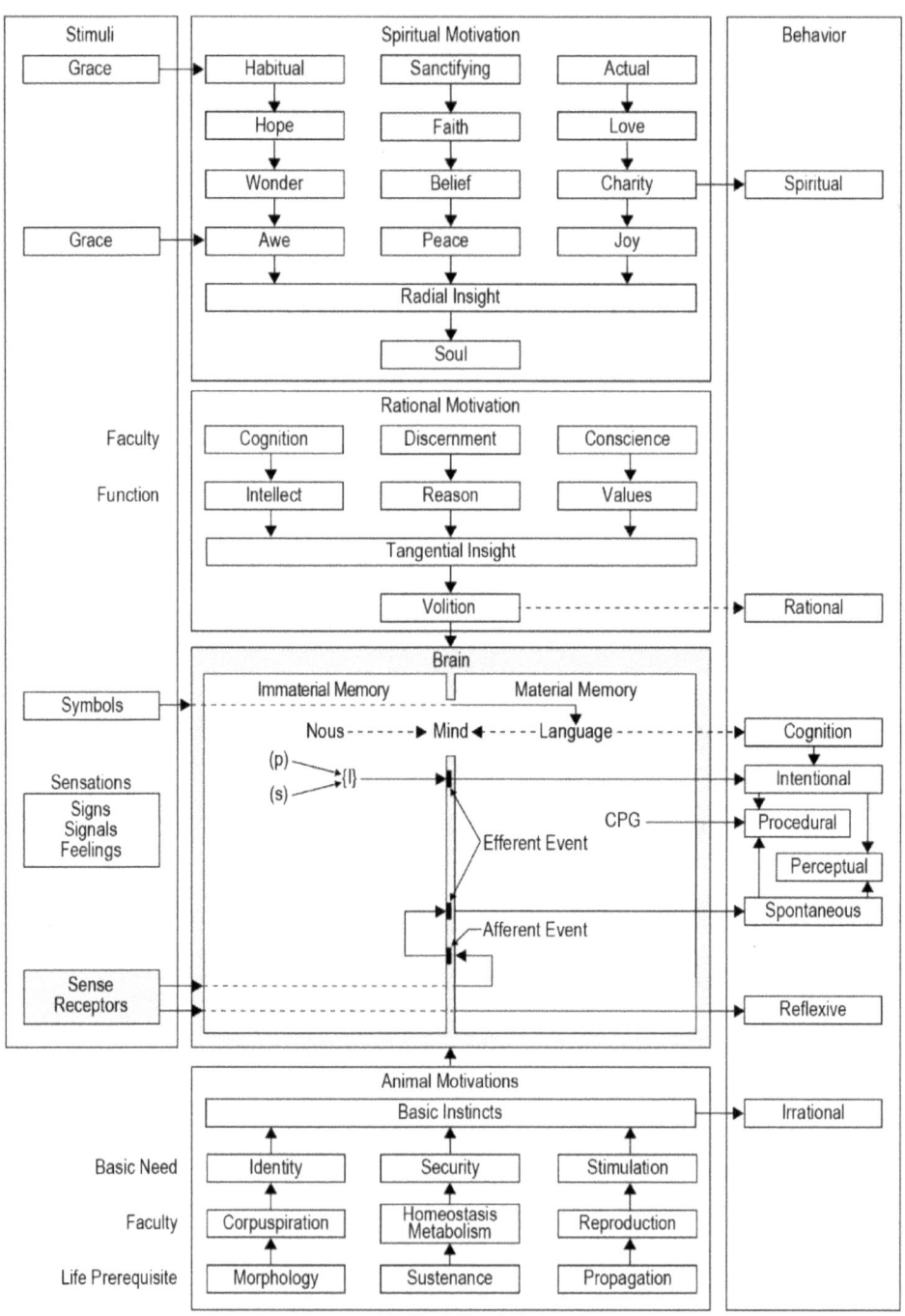

Figure 15. Spiritual Behavior

We are all familiar with our animal desires and the rewards they bring such as pleasure, fame, and power. The feelings they evoke are persuasive and hard to resist. They arise directly from the material component of our being and the needs for security, stimulation, and identity. Divine grace provides more subtle rewards like awe, peace, and joy. Unlike the animal rewards that are transitory, grace is cumulative and enduring; the more experienced, the more it is desired. Because the feelings that grace induces are subtle, they are difficult to recognize. But the rewards of grace are enduring, and once recognized and put into practice, they cumulate and intensify. Rejecting grace for power, fame, and pleasure leads to the enduring and detrimental states of ennui, anxiety, and despair.

God's grace works through body, mind, and soul. Habitual grace works through the body to induce wonder. Sanctifying grace works through the soul to induce justification. Actual grace works through the mind to induce charity. Tangential insight is composed of intellect, reason, and value; radial insight is composed of wonder, faith, and love. Radial insight strengthens one's rational motivation to do good by strengthening values. Radial insight modulates rational motivation through its effect on the value element of tangential insight. Whereas tangential insight provides the means for moderating our basic instincts, it is radial insight that provides the potential power for influencing rational behavior toward the good.

1. Sanctifying Grace

Grace is a gift given to us at baptism to aid us to seek justification through application of the theological virtues of faith, hope, and charity. Whereas habitual grace induces awe when we wonder upward toward God's creation, sanctifying grace induces a feeling of peace when we look inward for God in our own souls. I am a Roman Catholic, and in our religion, sanctifying grace is associated with the sacraments. I thought about calling this form of grace sacramental grace; the sacraments are the essence of my faith. Since Roman Catholicism is the only religion I know, and the only religion that I practice, I believe and receive the sacraments as the fonts of sanctifying grace. I know however that the sacraments do not alone induce justification, they merely help by strengthening o resolved to love God.

The sacraments lead into a discussion of religion, and I do not want to go there because I am not qualified. Where I prefer to go is in the basics of life, how to lead a good life, a holy one but one that avoids the bad and does good. And with that requirement, I can see no other way than to live in accordance with Judea-Christian principles. And you do not need to be a Catholic to live in accordance with Judea-Christian principles!

When I include sanctifying grace as one of the elements of radial insight, I do not exclude all those who are not baptized Roman Catholic, because I believe that the other two forms of grace can be sufficient for justification. My thesis is not meant to be an argument for or against any religion; it is meant to support the belief in God; it is as an argument aimed at the non-believers, the materialist, skeptics, and cynic. Consequently, I include under a sanctifying umbrella all who follow the Judea-Christian principles and profess a belief in God no matter how they worship.

Piety is the main psychical response to sanctifying grace. Piety is our interaction with God, both formally through discernment, prayer, and worship, but especially in one's acquiescence to the will of God, a passive element of the formation of the human soul. Piety is not just how one relates to God but is also how one divines God's will. God asks us to look inward to deny evil and to think compassionately. God asks us to look outward

171

with discernment at evil and with the awareness of the needs of others. Failure to respond to sanctifying grace results in feelings of anxiety. Unlike wonder with which we focus our mind outward on God's gifts of nature and life, piety is the focus of our mind inward on God's presence in our own lives.

By piety I mean more than the formal worship associated with religion and church. Piety to me has a broader meaning; it is primarily the application of faith to one's everyday life and when we add prayer tour daily practice, peace enhances. Hence, I write, the psychical state infused by God through reception of the sacraments appears in us as a mental state called *piety*. We accept this gift through actions such as worship and prayer, which when accomplished experiences a feeling of *peace*. That said, I am not claiming that Roman Catholicism is the only way to practice piety and find piece, "sanctifying grace" is my personal approach. I believe every human can find God in their unique way to find peace. But we must first accept and acknowledge the peace associated with sanctifying grace; to find God within one's own soul.

Peace is a feeling, not unlike awe, but minus the tinge of fear. It is the absence of fear, anxiety, worries, or guilt. It is the feeling one gets from a phone call that informs you that a loved one made it home safely. It is the feeling one gets when a medical report for you or a loved is one free of cancer. For me personally, peace is the same beneficial feeling as awe but tinged with comfort rather than fear.

2. Habitual Grace

Habitual grace is innate and accessible to every sapient human. Wonder is its most obvious manifestation. Awe is its most obvious reward. Grace is a psychical state of being, infused by God, that permanently inheres in the body and induces justification (the positive development of personhood or, traditionally our path from original sin to Godliness) by orientating one's life toward Godliness. In thinking about the meaning of life, I often wondered what the nature of the permanent grace that God gave felt like. I know it was not joy, which was related to relationships of love, it was not peace for that was related to acts of belief. And it certainly was not pride, pleasure, satisfaction, or any number of subjective experiences, they were all transitory, come and gone, and often led one away from God not toward Him. I do not know what theologians believe, but here is my own personal view.

Wonder is an innate gift that creates hope by finding God in the miracles that surround us, in nature, in the works of man, and in other people. It is through wonder that we find the purpose in life and in finding the purpose we can find its meaning. Only a life that has purpose and meaning has hope. Wonder is the main behavioral response to habitual grace. When the mind frees itself from the material needs and focuses on God's creations, it wonders. We wonder when we study, contemplate, and discover God's gifts of the universe, life, beauty, and truth. Wonder frees the mind from self-interest and enhances spiritual behavior.

There are times that wonder leads to a sudden discovery of new knowledge, a so called "eureka moment." These moments can be the non-psychical moments of discovery such as that of Einstein, Archimedes, and innumerable other scientists, technologists, inventors, and peacemakers that have resulted in countless examples resulting in new scientific, technological, and social actualities. For most, the eureka-moment often occurs when we grasp scientific or mathematical concepts. Or our eureka-moments can be psychical such as in the case of those well-documented examples associated with the saints. The transcendental examples are usually associated with the act of discovery; the psychical examples are usually associated with the act of wonder, and 'to wonder' is to accept God's habitual grace and the feeling of *awe*.

Awe is a feeling of surprise caused by something beautiful, unexpected, unfamiliar, inexplicable, and often tinged with reverence or even a bit of fear. Awe is a positive feeling in that it results from a specific experience. Awe is finding the solution to a differential equation. Awe is the conception and completion of a painting that *works*. Awe is witnessing a dew drop on a daylily. Awe is holding a baby son or daughter in your arms for the first time. Awe is the first glimpse of the Grand Canyon.

Without wonder and awe in one's life, without responding to God's habitual grace, one is in danger of lapsing into ennui. The feeling of awe each time we experience a moment of wonder at God's miracles: a clear nighttime star filled sky, the deafening silence alone at the top on a mountain; the quiet roar of the surf alone at the ocean; a baby's soft warmth in the stillness of the night. How could anyone listen to Ellina Baranga and Anna Trebenko singing Ofenbach's Baracola and think that neurons can create such beauty? To wonder is to take the first step toward hope.

3. Actual Grace

We experience joy when we practice charity by the sacrifice of our needs for that of another. Sacrifice is the psychical state temporarily infused by God to enlighten the mind or strengthen the will to perform actions that increase one's justification; it appears in us as a mental state signified as *love*. Love induces a wide variety of behaviors, the main characteristic of which is sacrifice. True sacrifice rewards us with a feeling of *joy*. *Sufficient Grace* is Actual Grace that we may or may not accept. *Efficient or Efficacious Grace* is Actual Grace accepted and acted upon through sacrifice and rewarded with a feeling of joy. *Sanating Grace* is the Divine grace in it function of healing, in human nature, the ravages of sin, original and personal.

Actual grace provides the joy we experience when we exercise the virtue of charity. Compassion leads to action that rewards us with a feeling of joy. By action I mean more than the formal action associated with community and service. Action to me has a broader connotation; action is the dynamic application of what we learn from wonder and piety. It is putting God's plan in action. It includes sacrifice, consideration, humility, charity, and love. Christ preached just such a way of life and that is why I know He is humanity's personal God. Failure to respond to actual grace results in feelings of despair.

Actual grace is the emotion that each human receives when he or she loves. Joy is the emotion God created to reward a person's act of love. When loved, a one-way action generates a false grace that we call pride. Joy is present in the action of love given to another and enhanced when love is mutual and triply enhanced when love is communal. The essence of Christianity is love, the essence of love is sacrifice, and all sacrifice is an act of charity. This joins love and charity, two words often interchanged as a theological virtue.

Keep in mind that this was my personal *philosophical* approach to finding meaning to the concept of grace and I will back off gladly anything I have written that may conflict with the theological view. However, I will not depart from the practice of wonder, belief, and sacrifice that has given me a lifetime of awe, peace, and joy. And it has oriented me towards faith, hope, and charity. But I will not deny any other person's form of worship as spurious if they: believe in the existence of God, and they follow the precepts of Judeo-Christian ethic.

Sacrifice is a word we think of to describe acts of bravery: the soldier jumping on a grenade to save the life of others; a firefighter entering a burning building to save a child; Christ on the cross. But sacrifice does not have to be of the dramatic kind; it can be associated with simple acts of everyday life: giving up one's place in line; visiting the elderly in an old people's home; or expressing any other small gesture with the following gestures: (i) to part with time and treasure, (ii) to resign oneself to a sacrificial offer; (iii) to offer your love. Most sacrifices are associated with the result of giving your time or your possessions. *Charity* is a more specific term than love, a more ambiguous term; it includes an act of sacrifice. However, whereas sacrifice connotes a loss of time or money, charity can be that, but it also has a positive side. It can be associated with feelings and positive feelings that can transform sacrifice into love.

C. Spiritual Motivation—Radial Insight

I presented animal motivation and rational motivation, now I introduce spiritual motivation. The basic instincts that produce the animal motivation is a faculty of the body, and if tangential insight that produces the rational motivation is a faculty of the mind; then radial insight that produces the spiritual motivation is a faculty of the soul. When one recognizes the existence of the soul, of a psychical element, or of the existence of God, the human mind generates a spiritual motivation that I refer to as *radial insight.* Insight is the ability to see into the true nature of things. And if one believes that there is a God, as my thesis does, then one must subscribe to the impact that God has on one's behavior. Radial insight is the personal understanding and exercise of transcendental resonance.

> Note 40: Transcendental resonance is the soul's direct interaction with the realm of potentiality. The interaction is an experience of an affection such as awe, peace, and joy, and that resonates with a subjective experience. For example, wonder (a subjective experience) resonates with a feeling (an affection) of awe.

Radial insight, the most important of the three motivational elements, is our ability to exercise our free will because radial insight provides the motivation for rejecting evil and for doing good. Tangential insight, the motivation of rational behavior, does not motivate directly, that is the role of the basic instincts and radial insight. The basic instinct motivates selfhood; the radial insight motivates personhood; the tangential insight moderates the other two motivators to allow volition to make a rational choice. Basic instinct influences negative volition, radial insight influences positive volition. I use the symbol {I} to represent volition in the corpus as the symbol for free will, the center of the soul, the integrity of which derives from the choices {**I**} makes.

When the basic instinct suggest vanity to satisfy a need for identity, the radial insight might counter with humility; when the need for security tempts with greed, radial insight can counter with moderation; when the need for simulation tempts with lust, radial insight can counter with decency. Just as basic instincts and tangential insight consist of three motivational elements, the three motivational elements of radial insight are: habitual, sanctifying, and actual grace.

D. Spiritual Stage

Like the other five stages, Noosgenesis consists of a seminal event, the creation era, and the objective. The other creation events were followed by an extended stasis period of growth. For example, the big bang was a seminal event that lasted about 380,000 years before stable hydrogen and helium formed and began the work of creating stars and galaxies. It took 100 million years before the first star began to shine. It was at that point that we can say that the universe was created. The seminal event lasted 380,000 years and the creation era lasted 100 million years at which time the objective (the universe) was reached, and the cosmological stasis period began. The spiritual stage differs from the first five stages in that it is not yet complete: we are in the midst of the era of creation for which the objective has not been reached or even known. I merely assume that the objective is the Omega point, the creation of world peace, and the end of ignorance and oppression.

This creation era, in which we find ourselves, is characterized by an increasing goodness that implies that reality was created with a purpose, a view that clashes with the positivist philosophy that there is no purpose in life, that life is nothing more than the result of a random interaction of chemicals. The meaning I have found in life, on the other hand, is fraught with hope built on purpose. I argue that the purpose of human life can be ascertained from the increasing goodness in the world, an increasing goodness that is a thinly veiled characteristic of human history.

PART 4— Spiritual Reality

Chapter 23—Hope

"For in this hope we were saved. Now hope that is seen is not hope. For who hopes for what he sees? But if we hope for what we do not see, we wait for it with patience." —Romans 8:24–25

Cynicism, the third argument of secular humanism, attacks the premise, God exists, with two questions: (i) the detrimental: if God exists why is there evil? and (ii), the purpose: why did God create us? Failure to explain the cynics questions diminishes the theists' hope. To answer the cynics' questions, I assume God's plan from three points of view: (i) the mystery of birth; (ii) the mystery of evil; and (iii) the mystery of death. God gives hope as an incentive to know, love, and serve Him. How? What hope? The theist answer is—we hope for eternal life. As stated in the previous chapter, I have a rational faith that impelled me to search for a plausible alternative to a "God did it" answer with which to argue the existence of God. This stance compels me to contemplate the nature of hope—be it eternal damnation in Hell, or joyful ecstasy in Heaven?

If one argues against the case against cynicism, one must first address the existence of evil. Since God is omniscient, God knows that evil is possible, that it exists. Surely an omnipotent God could eliminate evil, so evil must have something to do with God's purpose and we find a purpose in the following description of how God eliminates evil.

A. Mystery of Birth—God's Purpose

In this chapter I take on the difficult task of presenting the teleological argument that reality was created with a purpose, a view that clashes with the positivist philosophy that there is no purpose in life, that life is nothing more than the result of a random interaction of chemicals. The meaning I have found in life, on the other hand, is fraught with hope built on purpose. I argue that the purpose of human life can be ascertained from the increasing goodness in the world, an increasing goodness that is a thinly veiled characteristic of human history

Finding the purpose of human life is summarized in the simple question: why are we here? God created us and must have had a reason. The Catholic Catechism reads:

"God, infinitely perfect and blessed in himself, in a plan of sheer goodness freely created man to make him share in his own blessed life. For this reason, at every time and every place, God draws close to man. He calls man to seek him, to know him, to love him with all his strength. He calls together all men, scattered and divided by sin, into unity of his family, the Church. To accomplish when the fullness of time had come, God sent his Son as redeemer and Savior. In the Son and Through him, he invites men to become, in the Holy Spirit, his adopted children and thus heirs of his blessed life." —Catechism of the Catholic Church, Prologue, Paragraph 1

This entry has been taught to by countless Catholic students as an answer to the questions like: why did God make us? Why are we here? For me, applying the light of reason results in a generation of more questions than answers in my mind. "Share in his own blessed life"? "God draws close to man?" What do they mean? What in my experience has demonstrated such? My faith has attempted to derive answers that are more understandable. And in doing so I have, in my own mind, been able to say I do not believe that God exists, I know that God exists. My rational approach to faith in a way is a curse, for it has robbed me of the true faith, the sacred belief without proof, unconditional, uncorruptible, imbedded in heart and not in the mind. So, when I recite the Apostle's creed and say:

> *"I believe in God, the Father Almighty, creator of heaven and earth: and in Jesus Christ His only Son, our Lord, Who was conceived by the Holy Spirit, born of the Virgin Mary, suffered under Pontius Pilate was crucified, died, and was buried. He descended into hell; on the third day He rose again from the dead; He ascended into Heaven, and is seated at the right hand of God the Father almighty; from there He will come to judge the living and the dead. I believe in the Holy Spirit, the Holy Catholic Church, the communion of saints, the forgiveness of sins, the resurrection of the body, and life everlasting. Amen."*

In that short paragraph I proclaim my belief in the religious mysteries: God, God-centered creation, Virgin birth, Resurrection, Catholicism, Penance, the resurrection, and Heaven because it is my hope for an afterlife and whether or not that comes about, that at the age of 91, will occur in the not distant future. However, this is a polemic and the argument above is not an effective argument with the secular humanists, so I use an argument that is more direct and a better challenge to the secular humanist. I am not doing theology; I am providing an argument. The following paragraph with a polemic in mind that can be used as an argument not a defined dogma of the Catholic Church.

God knew everything except one thing: His own perfection. Because perfection is relative term, it is ascertained only by comparison, so God, to know His perfection, created an imperfect world with the ability to become perfect. God created us to function as a mirror, one that increases in clarity with time until humanity reaches the Omega point, and the mirror of humanity reflects no further change. Then God knows His own perfection and achieves omniscience.

A perfect world is one in which there is no evil. However, the absence of evil is a necessary, but not sufficient condition. To reach perfection the world must reflect total good. We must not only reject evil but also actualize good. I practice my religion because it is the only hope for what lies beyond this life. I am a Roman Catholic; it is the only religion with which I am familiar. I was born into a non-practicing Catholic family and did not become a practicing Catholic until I married my beautiful wife. It was through the sacrament of matrimony that I learned the meaning of grace and the value of Catholicism and the beauty of the sacraments. It was through the application of wonder that I learned Christianity. I am a Catholic trying to be a Christian; being Catholic and Christian is not always the same thing, although the purpose of Catholicism is to create Christians. I am not claiming there is only one way to achieve justification. My purpose in creating this thesis is to provide theists with a weapon in the form of a polemic, to counter the arguments of the unbelievers, not to elevate the claims of any one religion. I am only familiar with Catholicism but understand the simple purpose of Christianity. The essence of Christianity is love, and the essence of love is sacrifice.

> Note 41: Justification is the capability God has given humans with which to achieve righteousness through grace.

B. The Mystery of Evil—Cynicism

I defined my reasons for writing this treatise: to contemplate the meaning of life; and to provide an argument for theists to counter the arguments of non-believers regarding the existence of God. I defined the non-believers as secular humanists that includes materialists, skeptics, and cynics. Earlier, I pointed out that the materialist believes that God does not exist because matter and energy alone define reality and since the laws of physics define matter and energy, there is no reason to invoke a supernatural agency. The skeptic believes that God does not exist because there is no proof. The cynic believes that God does not exist because the presence of evil and suffering argues against a benevolent supreme being.

The cynic presents a tougher problem for theists; finding an answer to the question: "Why does God allow evil?" This question has been around for centuries. It is the leading question of philosophical cynicism since the Enlightenment. Consider this from:

[Pierre Bayle, Historical and Critical Dictionary]

"God is either willing to remove evil and cannot; or he can and is unwilling or he is either willing nor able to do so; or else he is both willing and able. If he is willing and not able; he must then be weak, which cannot be affirmed of God. If he is able and not willing; he must be envious, which is contrary to the nature of God. If he is neither willing nor able, he must be both envious and weak, and consequently not be God. If he is both willing and able—the only possibility that agrees with the nature of God—then were does evil come from?" —Neiman, Susan, quoted in Evil in Modern Thought: An Alternative History of Philosophy, Princeton University Press, pg.118

In any argument with a cynic, theists must deal with the question: "If there is a God and God is good, why did He create a world that is so evil?" The presence of evil in the world has always been a mystery for philosophers. The primary way for theologians to deal with evil has always been through religion. However, despite man's continuing supplication to a creative power, evil continues to exist. The religious answer is God does not create evil; persons do! This answer does not convince the cynic because the rejoinder usually is, but why does God allow people to be evil? And then we theists must deal with this from the Old Testament, *"I form the light, and create darkness, I make peace and create evil: I the Lord that do all these things."* (Isaiah 45:7)

In the past, philosophers addressed the question of why God creates evil—even when it is stated in the Old Testament—is to point out that the Bible does not mean personal evil. Their argument is to point out that evil is either associated with pain or with fault. I contend that: to deal with Bayle's challenge and the Old Testament, and to counter the cynic's arguments we need a practical answer that seems plausible. To find an answer, I first exam what we mean by the word "evil", unlike the philosopher's two modalities of evil, I contend that evil exist in three modalities: (i) personal evil (sin); (ii) social evil (injustice such as bigotry, slavery, tyranny) and (iii) natural evil (tragedies such as earthquakes, tornadoes, pandemics). The question now is: why does God allow: sin, injustice, and tragedy? And like the philosopher's solution, I believe that God does not create personal (fault) evil. I do not know why God allows evil, but I have arrived at an answer to counter the cynic's argument that seems possible and plausible. I am not a theologian but can read the dogmas and I cannot find in what follows anything that Roman Catholic dogma prohibits.

I begin by addressing personal evil. I address social evil and natural evil separately. Let us first consider the existence of personal evil. God is the realm of possibility, meaning that all things that are possible reside in the Mind of God. Evil is a possibility. However, a possibility is not evil, acts are, and until actualized and in the case of personal evil, it is the person that actualizes the possibility (gives in to temptation) that creates the evil, in the form of a deliberate sin. God's desire is to eliminate the possibility of evil; therefore, it must first be actualized in a way separated from God's Being so that evil in no way exists in Him. And if evil remains a possibility and not an actuality, it is not evil but is still a problem. Consequently, the method God chose to eliminate the possibility of evil was to create a scenario whereby evil could be instantiated in the mind of a human and eliminated bit by bit apart from Himself. Hence, the creation of human beings; God created us to eliminate evil.

If everything that is possible exists in the infinite nothingness, then any evil that is possible exists there. Since the infinite nothingness is the Mind of God, God's dilemma then is what to do about the possibility of evil, how to get rid of it? Perhaps He created humans to eliminate evil. Such a task necessarily requires the possession of free will to make choices. We eliminate evil from the realm of possibility whenever we reject evil and choose the good when confronted with a choice between the two possibilities.

God provided an example of temptation in the scriptures when in the desert when after forty days of hunger the Devil tempted Jesus three times, and He refused the temptation each time thus setting an example for humans to follow when tempted with a choice. For example, because of the way God created humans—animals with free will—possibility and its temptation of adultery becomes sin only if actualized. "Imagining the possibility" may be an impurity but is not evil. If a pair of humans, when confronted with the temptation to commit adultery, choose not to do so, they prevent the actualization of evil and effectively eliminate that specific possibility. Hence, a finite bit of a possibility of evil disappears from the Mind of God. If they choose adultery, it will appear in each successive parallel world until they choose for good, at which time they eliminate the possibility of that sin from the Mind of God.

I can imagine the scrupulous jumping up and down and waving their hands in vitriolic disbelief *when they read that God created sin.* No, I am not saying God created sin, I am merely presenting the question. Creation is an act of actualization of a possibility and an act is not evil until a human actualizes it; and in the case of personal evil, it is the person that actualizes the possibility. In this way we give meaning to the statement, "God does not create sin, humans do."

If humans are to make choices to eliminate evil, we must have free will. Free will is based on contingency, a contingency provided by the holonomic effect of evil. Thus, earthquakes necessarily occur because of the contingency required to make free will was built into the holonomic mechanism. I contend that God created us to eliminate evil. What I am about to write is to walk through an intellectual minefield, especially in the guise of a practicing Roman Catholic. I do not know God's mind, but to my mind what I write seems plausible and non-heretical since I am not claiming revealed truth but am creating a counter argument to the cynic the use the existence of evil to deny the existence of God.

1. Natural Evil

Natural evil is mostly a problem of nature straying from the normal, from an average. Hence, living things must suffer: (i) from weather effects (hurricanes, tornadoes, lightning, floods, fires, droughts, extreme temperature); (ii) from biological effects (disease, famine, handicaps); and from physiological effects: (addiction, mental disorders). For humans to eliminate evil, they must possess free will and exposure to choices. Humans, made in the image of God, possess a psychical component, free will, which has the capability, like God, for making choices. Choice results from contingency purposefully built into the morphonomic mechanism.. The contingency built into the holonomic mechanism is instantiated as f-noumena and stored in the pneuma. Consequently, to acquire free will, we must deal with disasters.

2. Social Evil

Behavior that impacts social groups such as families, political units; ethnic groups, and others. Social evil includes crime, war, slavery, abortion, terrorism, tyranny, propaganda, genocide, bigotry, and racism. In contemplating God's reason for creating us, I arrived at two possibilities: (i) to eliminate evil and (ii) to demonstrate His omniscience. I attempted an answer to the first question, "Why did God create a world in which so much tragedy, sin, and injustice exist?"

3. Personal Evil

Let us now consider the existence of personal evil, sin. God is the realm of possibility, meaning that all things that are possible inhere in the Mind of God. Because evil is a possibility, it exists in the realm of possibility, the Mind of God. However, an act is not evil until a possibility is instantiated, and humans actually do it. In the case of personal evil, God separates Himself from evil by actualizing a specific possibility that becomes a p-noumena in the form of a temptation, but it is not an evil until a person actualizes it by acting on the temptation it in the form of a deliberate sin. God eliminates the possibility of evil by instantiating it apart from His Being so that evil in no way exists in Him. And if a human rejects temptation, does not actualize it into reality, that sin forever ceases to exist even as a possibility.

Since the infinite nothingness is the Mind of God, God's dilemma then is what to do about the possibility of evil, how to get rid of it? Perhaps He created we humans to eliminate evil. Such a task necessarily requires the possession of free will to make choices. We eliminate evil from the realm of possibility whenever we reject the evil and choose the good when confronted with a choice between the two possibilities, So, here we have one reason God created humans.

C. The Mystery of Death—Heaven

God gives hope as an incentive to know, love, and serve Him. What hope? The Christian answer is eternal life. As stated in the previous chapter, I have a rational faith that impelled me to search for a plausible alternative to a "God did it" answer with which to argue the existence of God. This stance compels me to contemplate the nature of eternal life—be it eternal damnation in Hell, or joyful ecstasy in Heaven. Ironically, the non-believer's main weapon in the battle for the minds of humanity, science, has provided an excellent weapon for we theists called the *Many Worlds Interpretation* of the Schrödinger wave equation that implies the existence of parallel worlds. We theists can use the theory of parallel worlds to answer the cynic's most puzzling questions.

The parallel world hypothesis allows a plausible answer for mysteries such as Heaven/Hell; the existence of evil; original sin, conscience, salvation of souls, hope for those living life in suffering, and the efficacy of the sacraments. The parallel worlds hypothesis is useful for addressing the challenges of the cynic, although I dismissed the challenges of the skeptic earlier by arguing that assuming God exists has as much forensic strength as the skeptic's assumption that God does not exists. I will venture beyond equivalence and end this Part with a logical proof that God exists using the model of dual reality. In the meantime, I address the cynic from the theological virtue of hope.

God armed each of us with His grace in the battle with evil. How are we to conduct God's purpose when our lives are so short? What is the nature of Heaven and Hell, the incentives God created to motivate our efforts? Is everyone saved from eternal punishment? The answers I have arrived at have given me unconditional hope and faith and the desire to practice charity.

We find heaven and hell right here on earth. When we act in God's grace inducing awe, joy, or peace, we are experiencing a wisp of heaven. When we disregard God's grace and experience ennui, anxiety, or despair we are experiencing a wisp of hell. These psychical emotions, unlike our physical and mental sensations, are cumulative. We, in our daily actions, choose heaven or Hell.

So, what do I say to a secular humanist when challenged to describe Heaven or hell? I do not know for certain what they are. But I can imagine what I would prefer them to be. Being well into my ninety-plus years of age, each day heaven becomes an especially important object of contemplation. I decided that since I have never seen a description of Heaven that is both imaginable and plausible and accounts for that other possibility, we call Hell, I stopped guessing what Heaven and Hell are like and began to imagine what I would like them to be. It is easier to describe my Heaven than my Hell because fortunately this lifetime was closer to Heaven than to a Hell.

So, here is what I would like Heaven to be like. My mother and father will be in the same age relationship to me as they were this time around. They will not be teenagers, and they will not be aged; they will be my mother and father. So will my siblings, my children, their children and all the people that I have known will be there just as they are or were in this lifetime. Yes, there will be the same animals, flowers, oceans, stars, rocks and all the things I have experienced in this lifetime. I will fall in love again with the same beautiful woman and live an entire married life immersed in romance, good humor, and friendship. My Heavenly life will be filled with the same or more of the laughter, wonder, love, joy, fun, peace, nostalgia, and belief that has filled this life. I will hit a baseball again; I will hear La Boehme for the first time again; I will sing babies to sleep in the middle of a quiet night again. I will eat peanuts, smell roses, hear a whippoorwill, see the ocean for the first time; see Broadway musicals, watch my children graduate, be married to my wife, and have the same children again. That is my vision of what Heaven will be like.

On the Hell side, just as there is in this lifetime, there will be diseases, earthquakes, pandemics, floods, and all sorts of natural evil. There will be wars, bigotry, injustice, tyranny, poverty, and all sorts of social evil. There will be sin, crime, addiction, affliction, and all sorts of personal evil. However, in our next life each of these evils will be palpably diminished. Since I have had a minimum of disappointments in this life, I cannot describe a vision of a personal Hell, but it would consist of far too many regrets and sins, none of which I care to share. But if I have confessed those sins, transgressions, and regrets, then they will not happen in the Heaven that is my next lifetime because I will enter it with a more effective conscience, more moral certitude, and more open to God's grace.

So, my Heaven and Hell would look like my present life except there will be fewer regrets and sins; it would be palpably better. Think of the movie *Ground Hog Day* in which the actor Bill Murray plays a weather forecaster who repeatedly wakes up on the same day, but each new repeat, he alters his behavior for the better and experiences more joy. Each new life would be closer to Heaven and farther from Hell until I and all the rest of humanity achieved that goal, and we reach the fullness of the Mystical Body of Christ.

I am not saying with certainty that my view of heaven conforms to a theological view of Heaven in accordance with scripture and the "defined dogma" of the Roman Catholic Church, it is merely what I want Heaven to be. On the other hand, for those that have not lived a full lifetime of wonder, peace, and joy, and those now suffering in a life that seems like Hell, my view provides a possibility that describes how the unfortunate will eventually escape their individual Hell. Yes, all souls will be saved. In the meantime, we are making our way through our personnel Purgatories in which we too often make the wrong choices by failing to respond to God's grace. Eventually we will all escape our personal Hells and arrive at that perfect world we call Heaven. How would this sort of Heaven/Hell come about? There is a scientific solution for my hope, namely, mentioned previously as the Many Worlds Interpretation of the Schrödinger wave equation implies the existence of parallel worlds. The first thing that comes to mind for some that hear of the parallel world idea is reincarnation. Living a parallel life as yourself is not reincarnation; if the Many Worlds interpretation is real, we are living in one now.

Portions of our current life that imitate both Heaven and Hell are cumulative in radial time across worlds. Hence, each of us gets the opportunity to eliminate evil in successive lifetimes, so that each successive lifetime is *better* in terms of increasing good. Eventually every soul will be pure, every person will experience a life without regret. We will all have gotten to Heaven. Those who experience Hell on earth from tragedy, sin, and injustice will eventually live lives of awe, joy, and as their Hell decreases in radial time. Those who commit sin and injustice will accumulate Hell's worth of ennui, anxiety, and despair. The retrogression of increasing good in radial time, and here on earth in tangential time, will eliminate evil in all three forms—physical, social, and moral. I find theological support for my view in the following that states in Step 29 of Vatican's 2004 International Theological Commissions report on *"Human Persons Created in the Image of God"*

> *The central dogmas of the Christian faith imply that the body is an intrinsic part of the human person and thus participates in his being created in the image of God...The effects of the sacraments, though in themselves primarily psychical, are accomplished by means of perceptible material signs, which can only be received in and through the body. This shows that not only man's mind but also his body is redeemed. The body becomes a temple of the Holy Spirit. Finally, that the body belongs to the human person is inherent to the doctrine of the resurrection of the body at the end of time, which implies that man exists in eternity as a complete physical and psychical person.*

From this I infer that since our bodies are resurrected at the end of time to exist eternally as a complete physical and spiritual person, and since in this present state the body and soul are imperfect, and since it is through grace and the application of our free will that we are justified, it seems plausible that there be interim states in which the process of justification is completed; hence Purgatory.

D. Purgatory

In the New Testament, the clearest reference to purgatory comes in Matthew's Gospel (12:32), where Jesus states that *"whoever speaks against the Holy Spirit will not be forgiven, "either in this age or in the age to come"*—implying that sin is forgiven in the next life. Such scriptural references lead to the church's belief, stated in the Catechism of the Catholic Church, that:

> ***"All who die in God's grace and friendship, but still imperfectly purified, are indeed assured of their eternal salvation; but after death they undergo purification, so as to achieve the holiness necessary to enter the joy of heaven."*** —Catechism of the Catholic Church, No. 1030)

Chapter 24—Faith

"Belief is believing in God; faith is believing that God believes in you."
—Andre Dubus

Skepticism is the philosophical view that only reason can find absolute truth. Skepticism when applied to religion is disbelief and is translated by the atheist that since there is no proof, God does not exist. This dogmatic assertion distinguishes the atheist from agnostic who is not certain if God exists or not. My argument is with all non-believers including agnostics. My argument with the atheist is that the skeptic cannot prove that God does not exist, and the argument shifts to faith. Faith is believing without proof. Atheists contend that faith—belief without proof—is a senseless position. However, faith works both ways. Theists have faith without proof that God exists. On the other hand, atheists believe without proof that God does not exist. Hence atheists that believe without proof that God does not exist, necessarily have a type of faith. Better to be an agnostic.

However, my aim here is not to find proof of God, but merely to diminish the non-believer's use—in a debate— of science as a proof. Science does not prove that God does not exist; the faith that atheists fall back on is that science sufficiently describes what we can observe and there is no need for God. This atheistic fallback position is scientism, the idea that reason alone can find truth. We interpret this to mean that truth requires evidence that implies that the only way we can find truth is through science. However, believing that scientism is the only path to truth is itself an article of faith.

A. Countering Scientism—Plausibility

I argue that the premise 'God exists' has as much forensic weight as the premise 'God does not exist.' With the premise that God exists, I explain how God might create and sustain reality at an explicative level while what we experience, and science describes, is at a descriptive level. I have not found proof that God exists, rather, I believe my view levels the playing field by providing an alternative view of reality, and it strengthens one's faith in the existence of God with a deductive approach that shifts the argument from proof to plausibility. This diminishes *scientism* by demonstrating that science is not the only path to truth.

> Note 42: Scientism is the belief that only science can find the truth of reality, Science, by restricting explanation of reality to physical elements (space, time, matter, and energy), has not been able to explain psychical phenomena such as consciousness, qualia, universals, and the mind.

By separating my view from science's view without diminishing science, I argue that my view (the MDR) is a plausible alternative to science's common view of reality, hence, it counters scientism. In this way a theist is presented with an answer, other than "God-did-it," to the materialist's and skeptic's arguments.

The MDR is not a proof of God; it is simply a way to level the playing field between theists and non-believers by countering the contention that only science can find the truth. Finding an alternative view of reality reduces the stature of scientism by providing the theist side with a view of reality that is more comprehensive than the disparate theories of science. Intelligent Design proponents and other theists argue the improbability of life occurring stochastically along with the "God-did-it" tenet. My approach to countering the scientific view is to supplement the God-did-it tenet with a plausible argument as to **how** God did it. In other words, I am offering a plausible argument to show science is not the only path to an explanation of the nature of reality.

B. Countering Materialism—Dualism

Although most secular humanists believe that reality is solely material, there are those in the field who recognize a need for dualism to find truth of reality but have not found the simple way of describing it. Although science has done a credible job in dealing with physical elements, by restricting the explanation of reality to a materialistic view, science has been far less successful in dealing with the psychical phenomena that demands dualistic explanation of reality. That dualism exists is evident by considering the evolution of the universe.

For billions of years, the universe was solely materialistic, and science's description of reality was relevant. However, when the first complex multicellular organisms appeared, a new phenomenon (sentience), came along with them. Metazoans experienced sensations and sensations are non-material. Materialism cannot explain sentience; it can only describe it as an emergent property of the brain.

The *sentient stage* began then was followed by the sentient stasis period of about 600-million years ending about 50-thousand years ago (kya) when the first human mind appeared. Other than trace fossils, multicellular organisms did not appear in the fossil record until about 540 mya when paleontologists discovered fossils in what is known as the Cambrian Explosion. Multicellularity represents an enormous increase in material complexity giving it the appearance of a creation event that I refer to as somagenesis.

My view, based on a physically derived hylomorphic duality, provides a better argument than the materialist's contention that we can explain the nature of reality solely with the physical elements. The capability of hylomorphic dualism in addressing both the physical and the psychical produces a comprehensiveness that diminishes the view of scientism and hence diminishes the view of materialism.

C. Countering Skepticism—Plausible Proof

I counter scientism, without diminishing science, with a plausible dualistic view that provides the theist with something other than "God-did-it," as a rejoinder to the materialist's arguments. The premise 'God exists' has as much forensic weight as the premise 'God does not exist.' Although my view of reality may not be the abstract truth, it may be so sufficiently plausible that it could open doors for science to advance. This is especially true for the physical sciences that deal with the mind.

My view is not an abstract proof that God exists it merely hopes to strengthen one's faith in the existence of God with a deductive approach that shifts the argument from proof to plausibility. The explanation of how God creates and sustains, even if it does not pass a plausibility test, at least provides an argument that there is an alternative to the common view that in effect diminishes scientism as a sole path to truth. On the other hand, what I have proposed can be offered as a deductive proof.

D. Deductive Proof

Although I present my argument for the how God exists not as proof, but on the other hand, it is hard to ignore the use of the physical element, continuous space, in the following syllogism as a plausible deductive proof of God's existence:

Nothingness Exists / God is Nothingness / Therefore God Exists

Part 1 discusses the credibility of the first premise; nothing exists. I argued that infinite nothingness is the only way to describe the pre-universe that preceded the big bang. I argue the credibility of the second premise, that God is nothingness. with the following: nothingness is the substance of the pre-universe and is devoid of matter and is indescribable. Since the pre-universe is infinite, it contains all that is possible. Since nothing can be added or subtracted, it is unchangeable and formless, therefore immutable. Since the pre-universe existed before time, it is eternal. Since the pre-universe contains and can implement all possibilities, it is omnipotent. Since the infinite nothingness, in the modality of pneuma, pervades each particle of matter in our physical universe and forms a seamless cloak with the infinite nothingness of the pre-universe, the infinite nothingness is omnipresent. Hence the infinite nothingness of the pre-universe is infinite, indescribable, immutable, eternal, omnipotent, and omnipresent. We need only to demonstrate that the pre-universe is omniscient to argue that the pre-universe is what we mean when we refer to the transcendent God. Until I demonstrate how the pre-universe is omniscient, I will assume that the pre-universe has all the attributes of the transcendent God. Then, in a sense, the infinite nothingness that pervades every s-point in our universe as pneuma is the Creator and, of course, the Intelligent Designer.

An example, of how the MDR can be applied: when for example the skeptic asks: how can anyone believe that Jesus turned water into wine when water is composed of hydrogen and oxygen and wine is composed of water and an extremely complex combination of other chemicals such as carbon organized as a plethora of hydrocarbons. If organized as we observe and science describes reality, then any reasonable mind would tend toward the impossibility of the miracle. However, if reality organizes at the ground state as I suggest, not of atoms and molecules, not of quarks and electrons but of s-points, then turning the configuration of s-points that manifests as water into a configuration that manifests as wine does not contradict any possibility associated with God's power, especially since in the model of dual reality God has 10^{43} increments in each second in which to affect the conversion. So, God, as Jesus, worked at the lower ground of reality that is composed of s-points to alter the local part of the cosmic s-frame to reconfigure the s-points that represent water molecules into a local reconfiguration that is wine. Remember the greatest miracle is a scientific one in which all the elements that constitute the universe appeared from an infinitesimal object—the singularity— in less than a second. Such a miracle turns the event of turning water into wine into a non-miracle.

In my view, I know Jesus is God. I know this because in my search for meaning, everything Jesus said rings true, and one of the things he said is: *I am the Son of God.* This is the foundation on which I build a faith of the mind instead of a faith of the heart. The faith of the mind is never a true faith because faith means belief without unconditional proof and faith of the mind is at most a search for proof, no matter how you frame it.

I developed my polemic view by "assuming that God exists" and operates at a deeper level of reality. This allowed me to describe a dualistic reality that is plausible and more comprehensive than the common view. Plausibility is derived from the dualistic view of space in which continuous space is viewed as psychical and discrete as physical. From this I presented a deductive proof of the existence of God. Having taken that leap of logic, I found another way to strengthen the existence of God with adjective proof based on Goodness.

Chapter 25—Increasing Goodness

My approach to the contention that goodness is increasing in the world is to identify the components of goodness and then examine how they have changed throughout history. Goodness is an abstraction that can be discerned by examining the evolution of the main aspects of human life, namely intellectual, social, and moral order and their consequent effect on the four levels of reality, objective, subjective, rational and transcendental. The principles on which intellectual order is based are technology and education; the principles of social order are law and order; and the principles of moral order are personhood and selfhood. The relationships among the various factors associated with increasing goodness are shown in figure 16.

As shown in the chart, the part of our destiny for which goodness can be observed are three faculties of human life: intellectual order, social order, and moral order. Intellectual and social order are composite; moral order derives from individual humans. For example, knowledge can induce good or evil behavior. Intellectual order consists of knowledge and technology; social order consists of law and justice; moral order consists of personhood and selfhood. Both intellectual and social order effect goodness indirectly through the impact on moral order. Goodness is generated by the summation of individual morality; morality is the direct measure of goodness. Morality, the individual order has a negative side that we call evil.

I contend that there is a historical destiny, the deep current in the affairs of humans that is independent of individuals and events. If not one way than another, history happens and happens toward the same end. The result is inevitable. Only the sequence and timing are different. I argue that historical destiny is driving humanity towards a perfect world by increasing the integrity of the noosphere through a positive increase of goodness and a decrease of evil. If God exists and I believe God does, and since my God is a triune God, it is the Holy Spirit that is the impetuous driving our destiny.

Given the history of the human race, with its wars, famines, plagues, social injustice, and immorality, finding an argument for the increase of goodness is a challenge. Especially since just virtually every improvement in a component of goodness has a downside. For example, the first tool, a good thing, soon became a weapon, a potential evil thing.

The opposite of goodness is evil. What I mean by the increase in goodness then is that which exceeds evil; hence any decrease in evil can be counted as an increase in goodness. Therefore, there are two possibilities for increasing goodness, by addition of the one or the reduction of the other. I address the three modalities of evil: natural evil (tragedy); social evil (injustice); and personal evil (sin) when I deal with question as to why a benevolent God allows evil in Part IV.

Here I deal with impact of evil on increasing goodness along with the positive approach of directly increasing goodness. I discussed goodness earlier at <7.D.d> where I pointed out that goodness is an abstraction that ties together diverse transcendental experiences such as beauty, truth, and justice. The transcendentals are components of goodness that are derived like beauty from transcendent or like justice and truth that are derived from social sources. Humans create things that are beautiful, but unlike truth and justice that are derived from human behavior, the noumena that is beauty is not changeable hence doesn't take part in the increase of goodness. Truth and justice can be changed as the sum of individual truths and instances of justice, to both the individual and the social. The individual metric of goodness is virtue, the application of faith, hope, and charity. To go along with the collective metric of goodness, the experiential increase of truth, justice and beauty. The path to the personal goal is justification; the path to the collective goal is the omega point, perfect world peace.

Transcendental reality is shown apart from the other three because it is purely immaterial whereas the other three levels of reality originate from a material base. Transcendental reality inheres in the Mind of God and in the goodness sphere provides beauty that is manifested through nature.

There is much in the transcendentals that is subjective. There can be disagreement about specific experiences so that what one person finds beautiful, truthful, or just, others may not. However, there is general agreement that if something is beautiful, happy, or just then the common transcendental that can be agreed upon is goodness. I know the word 'good' is an abstraction, but I assume that there is common understanding of what it means, and what it means implies a positive feeling, whatever that means to you. The transcendentals are good by definition; their innate goodness cannot be diminished. Transcendentals don't reside in the immaterial memory but seem to be apart from our minds, for example beauty inheres in Anna Netrebko singing Puccini's aria, *mio babmbino carra*; truth inheres is the observation that the world is not flat; justice inheres in the 13th amendment of the American constitution's abolishment of slavery.

Certainly, there are more goodness-evoking transcendentals such as equality, courage, and honesty that are not included. For this synopsis, justice, happiness, beauty, truth, and justice suffice to argue that goodness is increasing. More justice, more happiness, more beauty, more truth, means goodness is increasing. Of the intrinsic transcendental, happiness can be tied to technology, but it is far more difficult to deal with virtue; its treatment must wait until later.

A. Progress Principles

The three progress principles have shown discernible progress. Education progresses by expansion; technology progresses by invention; and polity progresses by the growth of democracy. Education, technology, and polity for which progress can be represented with a growth curve. The shape of their common progress curve is shown by a generic curve in fig. xx. This is the classic sigmoid growth curve in which a long period of minimal progress is interrupted by a relatively short period of rapid growth levelling off in return for the slow growth. To describe the progress, I select 3-factors on the curve: (i) the *seminal point* as the point in the stasis period from which rapid growth begins; (ii) the *expansion slope* as the duration of the rate of exponential growth, and (iii) *creation point* as the completion of the goal. These three factors will be linked to transcendentals to represent increasing goodness. The progress curves I present are literacy for education; life expectancy for technology; and polity by the Polity IV scale that measures the extent of democracy. We can link the goodness of transcendentals with the principles of progress to show how goodness has increased. The principles I have chosen are education, technology and polity.

1. Education

Education is a faculty; learning is its function. Its objective is the acquisition of knowledge. Since there are two forms of knowledge: (i) conceptual and (ii) perceptual, there are two methods of learning: (i) conceptual knowledge is acquired through academics; (ii) perceptual learning is acquired through training.

Academics are associated with the acquisition of conceptual knowledge, and most of what is learned through academics is in books, and books are filled with words, and words are discrete, academic learning enhances one's conceptual intellect. Conceptual knowledge includes mathematics, physics, chemistry, biology, philosophy, history, literature, etc. The effectiveness of acquiring conceptual knowledge depends on one's innate intelligence.

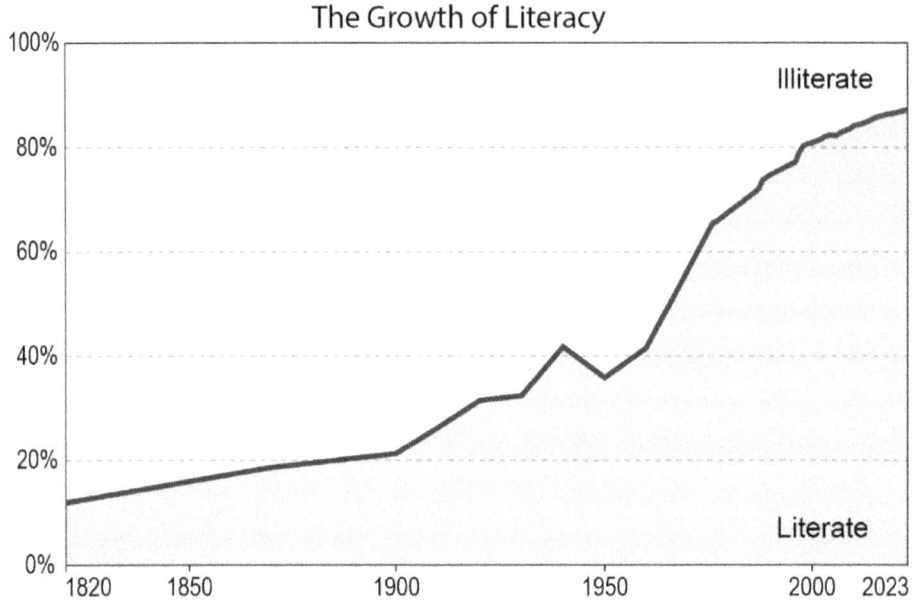

The Growth of Literacy

Data source: UNESCO (1957) and other sources, OurWorldinData.org/literacy | CC BY

Figure 16. World Literacy Rates

Training is associated with the acquisition of skill or "how to do" something. And there are many how-to-dos. How to dance, how to row a boat, how to bake a cake, play a musical instrument, paint a picture, install plumbing, shoot a basketball etc. Training is mostly associated with motor memory and since motion is continuous, training enhances one's perceptual intellect. The ease of acquiring a skill depends on the innate capability known as talent.

To argue the increasing goodness in this ocean of perceptual and conceptual information, we need to find the most pertinent progress principle, and for me, the one that stands out, one in which forms the foundation, without which conceptual and most perceptual education is impossible, is literacy, the capability to read and write. The accompanying chart depicts the progress of education based on literacy. The y-axis represents the percent of the world's population over the age of 15, that can read and write; the x-axis represents the years 1650 to 2023. The base progressed from 0 at the advent of reading and writing thousands of years ago to the 18% at the seminal point in 1850, to the 87% today. The single most pertinent cause for the exponential increase in education is the introduction of compulsory elementary education throughout the countries of the world. Although there are some skills that can be acquired without being literate, it is essential for conceptual intellect. Therefore, I argue that the education has increased as the result of increase in literacy.

190

2. Technology

Technology is the use and modification of matter and energy in an attempt to increase the quality of life. I hedged my bet with the word 'attempt' because the use of weapons, one of the products of technology, more often than not decrease the quality of life, and in effect, not just the quality of life but life itself. Although in the history of humanity, warfare and its need for weapons has often resulted in the advancement of technology. Science also plays a part in the advance of technology, but not with as much as science is credited. Long before there was science there was technology. In fact, technology has done more for science than science has done for technology. Science could not have diverged from philosophy without the tools for observation and measurement.

Technology covers a broad range of human endeavors; it is an important impetus for the success of the following aspects of the quality of life: (i) sustenance—the production, preparation, and preservation of food; (ii) safety—protection from danger and the elements, (iii) the social aspect—the ability to congregate, interact, and when necessary to migrate; and (iv) intellectual aspect—the desire to know, create, and enjoy life. The underlying faculties that are related to some or all of the aspects are construction, transportation, productivity, and communication. For example, the capability to produce more food (i) and knowledge (iv) is a good thing; the capability for better communication provides more protection against dangers (ii); and improved transportation increases the ability to congregate (iii).

Of the progress principles, technology's progress is the most discernable. There is a record of the evolution of technology from discovery of stone tools 3.5 mya to the Internet of the 21st Century. One need not go farther to contend that technology has made great progress. However, the progress of technology was not one long uniformly increasing progression. Instead, technology formed a minimally increasing progression until the 19th century when it virtually exploded with what is called the first industrial revolution that increased the rate of progress that still continues today.

Technology began with a discovery, probably a simple tool like a stick. But technology tends to stagnate unless an application is found for the discovery. Who knows how long men held sticks before one of them thought to use it—perhaps to reach a piece of fruit hanging from a limb? Technology took a small step forward the day of that first application. Other applications followed. The stick was used to dig for roots or to ward off and even kill animals. Improvements soon followed. Perhaps a stick with a fork on the end was used to pick the fruit. A heavier stick to ward off animals, and a stick with a sharpened tip was used to dig for roots. Thus, we see discovery induces application that induces improvement and often in applications and improvements new discoveries are made.

Technology covers a broad range of human endeavors; it is an important impetus of success for following aspects of the quality of life: (i) sustenance—the production, preparation, and preservation of food; (ii) safety—protection from danger and the elements, (iii) the social aspect—the ability to congregate, interact, and when necessary to migrate; and (iv) intellectual aspect—the desire to know, create, and enjoy life.

Discovery, application, and improvement enhance and induce one another and in this way, man weaves a technology with three elements as its strands. From this concept of technology, we can define a technologist as one whose prime task is to increase the level of technology by discovering better ways for man to interact with his environment and by finding new ways of applying and improving those discoveries. By technologist I mean not only the engineer but also the perceptive basket weavers, farmers, toolmakers, potters, and other types of craftsmen. Later when the Greeks invented science, technological advances resulted as a by-product of science's primary search for relationships and explanations of natural phenomena. Even later, technology was formalized by the profession of engineering.

Technology began when man discovered his first tool. Engineering came later. Engineering was the child of agriculture. The discovery of agriculture eliminated man's need to wander in search of food. He settled in towns and cities. And this permanence created a need for buildings, roads, and bridges. These structures gave rise to new problems calling for planning and for the application of experience, innovation, and accumulated knowledge. It was the formal assignment of a problem that distinguished the engineer from the craftsman. From the irrigation systems of Mesopotamia to the pyramids of Egypt, to the harbors and temples of Greece, engineering grew like the trunk of a tree, without branches, solving the same kinds of problems. This concentration of engineering on structures reached a pinnacle with the Romans.

For 1000 years the Romans ruled the western world. To maintain their vast empire, they built an extensive system of roads. They supplied towns with water by aqueduct, erected magnificent buildings, devised central heating and indoor plumbing, and invented the arch and concrete. And performed many other engineering feats. They mastered the field of civil and structural engineering. In addition, their preoccupation with conquest led to a new branch of engineering.

To solve the problems of breaching the walls of fortifications, the Romans devised numerous "ingenious" devices called engines of war such as the catapult and the ballista. It is from these forerunners of the mechanical engineers, the Romans who worked on the engines of war, that the name engineer is derived.

Then, as the Roman era ended, the Imperial system was decentralized into the feudal system, so too was engineering. Engineers became scattered and less concentrated than when they marched with the Roman Legions. Technology slowed down—the emphasis shifted from engineering of structures to the engineering of machines. The Engineer's main quest throughout the Middle Ages was for better ways to use the only sources of power available—wind, water, and animals. And many were found. But even though structural engineering found a renewed burst of energy with the building of medieval forts and Gothic cathedrals in the 12th and 13th centuries, engineering did not expand. Even during the Renaissance and the age of Reason man's energies were devoted to science and art. But in the middle of the 18th century engineering burst from its main trunk into the many branches we know today.

As the discovery of Agriculture gave root to the trunk of engineering, the use of power was the bud from which its branches burst. The invention of the steam engine by Newcomen in 1723 and its improvement in by Watt in 1759 increased the power available to man thus giving impetus to the greater use of machines. This ushered in the Industrial Revolution. Each new discovery or invention led to new applications and improvements calling for engineers to become specialists. Thus, the invention of the dynamo in 1867 along with Faraday's discoveries led to the use of electricity and the need for electrical engineers. Increasing use of metals in machinery created not only the need for metallurgical engineers but also the need for mining engineers. In 1855, Le Blanc developed a commercial process for the production of $NaCo_3$—thus initiating the chemical industry with its need for chemical engineers. The internal combustion engine gave rise to automotive, petroleum, and aeronautical engineers. And so on into the 20th century with its many new discoveries, each of which opened new technologies. and the life expectancy began to grow.

Meanwhile during the 19th century there appeared another force that had a dramatic impact on technology, the force of capitalism that gave birth to the factory. Prior to then, the mechanism of production was the "domestic system." Entrepreneurs would accumulate money, buy raw materials, distribute them to craftsmen and find markets for the finished product. But now the steam engine provided a way to power many machines simultaneously. It became practical to bring many workers under one roof and the factory was born. And in the factory, the division of labor gave each worker a specialized task at which he could become proficient. The result was a greatly increasing impact on human sustenance and the increased beauty and happiness that it creates.

Technology impacts many facets of human life such as communication, transportation, computation, materials, medicine, energy, tools, weapons, agriculture and others, which selectively have an impact on the availability of beauty, and the increase in sustenance. Technology impacts goodness in two ways: (i) the experience of beauty, and (ii) the individual's need to fulfill sustenance. Technology's impact on fulfilling the individual's need for substance was a driving force in the progress of the polity principle.

3. Polity

I use the word *Polity* to mean the way that humans are organized and governed as social units. Polity is a faculty; its functions are governance and human rights. Since the first pre-human walked on earth the driving force in human organization has been, and in some ways still is, the need to fulfill the basic human needs for identity, security, and stimulation. Polity consists of two principles, liberty and equality, that impact goodness through the same transcendental of justice. Organization of the social unit, from family to clan-sized social units, was driven by the need for security and held together by the need for identity. The size of social units advanced from that of a clan, a loose organization of related families, to that of a tribe, or a similar social unit with a more well-defined polity.

Through the ages polity, in with the principles of liberty and equality, were present in a wide range of forms. However, it can be argued that the form with the greatest success in producing liberty and equality is democracy. Progress in polity is associated with the increase of democratic countries (figure 17). This progress along with the associated equality and liberty satisfies the need for personal identity and happiness. The basic polity model that most likely was established millions of years earlier by the precursors of the genus Homo is based on that of the traditional hunter-gatherer family in which the male leads, provides, and protects; while the female is caretaker, gathers, and propagates. The model has always been hierarchal in which typically there was an alpha-male that governed while the beta-males were the hunters and protectors, and the females did the gathering.

Eventually as the social unit expanded and got sufficiently large, while the roles of the alpha and the beta males remained the same, a third role appeared, the gamma-males. The gamma-males consisting of elder males that provided advice and males with special skills that filled support roles such as the makers of tools and eventually evolved into the bourgeoisie. While the alpha and beta roles persisted for most of pre-human to human history, the discovery of agriculture generated a proliferation of gamma-roles. The gamma-role functioned as the root of an expanding tree of information, commerce, and diversion roles in which over time developed numerous branches. The information role diverged into science, mathematics, philosophy, history, literature technology and religion. The commerce role diverged to form traders, sellers, explorers, inventors, and craftsmen. The diversion role diverged into entertainers such as storytellers, athletes, actors, and musicians. The alpha-beta-gamma model was the blueprint from which all subsequent polities evolved. The three-role polity existed until the social unit outgrew the constraints that held the unit together and produced too many innate alpha males.

Throughout most of human history, governance centered on an alpha-male. As polity evolved with a sequence of forms based on number of members through the family, clan, tribe, city-state, province, nations, and empire, we find the power in the hands of a patriarch, chief, lord, monarch, president, dictator, and emperor.

Some beta males still provided advice to ruler while other betas filled alpha roles in the lower levels of the social unit as lords, generals, dukes, governors, etc. Power cascaded downward through as many levels of power in the social unit, stopping only when it arrived at the deltas who were completely powerless. The beta's sustenance role diversified and evolved from hunter to include shepherd, fisherman, and farmer; the beta's protector role evolved from warrior to soldier. The gamma's support role evolved from tool maker to include trader, explorer, artist, craftsman, builder, and many other roles; the beta's advisor role evolved through elders, shaman, medicine man, priest, teacher, artist, politician, scientist, etc. The diversity of roles began after the domestication of animals and agriculture took root, then the numbers in the gamma role mirrored the growth rate of technology. The αβγδ-model, with few exceptions, persisted throughout history, setting a hierarchal pattern that has manifested itself as aristocracies and other class systems. Although class-based polities can serve the principle of liberty, they by their nature fail the principle of equality.

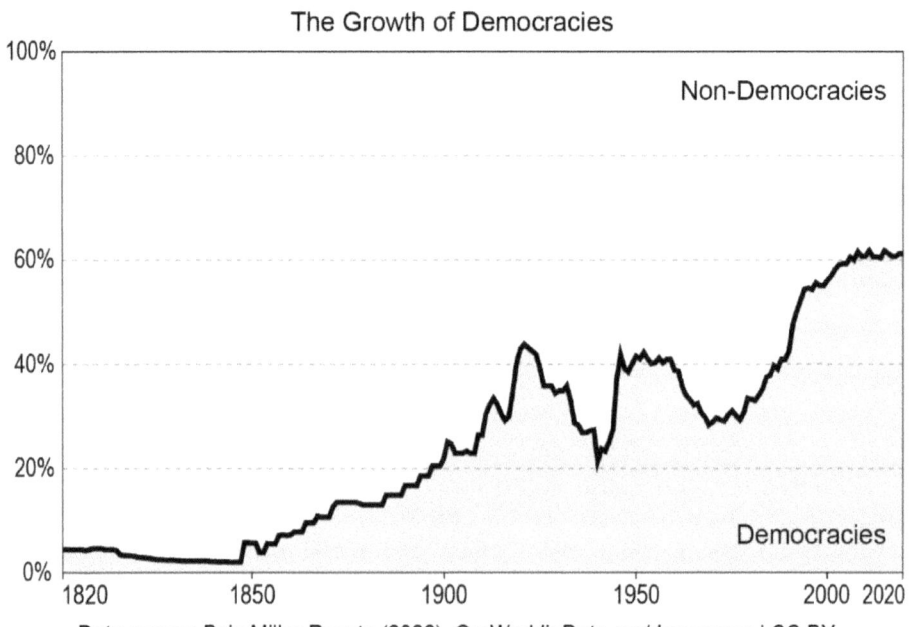

Figure 17. Growth of Democracies

Meanwhile the beta's sustenance role diversified and evolved from hunter to include shepherd, fisherman, and farmer; the beta's protector role evolved from warrior to soldier The gamma's support role evolved from tool maker, trader, explorer, artist, craftsman, builder, and many other roles; the advisor role evolved through elders, shaman, medicine man, priest, teacher, artist, politician, scientist, etc. The diversity of roles began after the domestication of animals and agriculture took root, then the numbers in the gamma role mirrored the growth rate of technology

The betta advice role was the group from which bishops, cardinals, popes, mullahs, formed a strong parallel polity. Meanwhile this basic model in which males and females continue to provide gender specific roles has been maintained as the size of the social unit increased from: clans, to tribes, to city-states, to provinces, to nations, to empires, and to other models in between. Not until well into the 20th century was this basic familial gender-specific model challenged so vociferously that even the concept of gender has been brought into question and women began to fill the traditional male roles. However, the biggest deviation the model was manifested when a number of highly capable women filled the alpha roles. Hence: Cleopatra, Elizabeth I, Victoria, Catherine the great and others. Eventually, with the establishment of cities, a fourth role appeared, the deltas. The deltas form the underclass consisting of those workers without rights, slaves, peasants, undesirables, and unemployed.

To show that polity and liberty impact on the increase of goodness, I link polity and liberty to the transcendent justice. The faculty that links polity-liberty principles with justice is human rights. that have always been control by the social unit's alpha male. For example, the King (alpha) determined the rights of the lords (betas), that determined the rights of the underclass (gammas). Until the Magna Carta in England, even the lords didn't have rights. It took another couple of centuries for England to introduce a full parliament that had power along with endowed rights.

The first human emerged from an anatomically modern human species—perhaps a H.sapien Idaltu– after tens of thousands of years, when the first sapien acquired the faculty of semantics, see <ch.21B>. Semantics allows a person to think, and thought is what defines the Homo sapien as a human. This progenitor of the human race was a part of a social unit of non-thinking pre-humans that had the same patriarchal polity for over a million year as migrating bands of nomadic hunter-gathers.

B. Transcendentals

There is much in the transcendentals that is subjective. There can be disagreement about specific experiences so that what one person finds beautiful, truthful, or just, others may not. However, there is general agreement that if something is beautiful, happy, or just then the common transcendental that can be agreed upon is goodness. I know the word 'good' is an abstraction, but I assume that there is common understanding of what it means, and what it means implies a positive feeling, whatever that means to you. The transcendentals are good by definition; their innate goodness cannot be diminished. Transcendentals don't reside in the immaterial memory but seem to be apart from our minds, for example beauty inheres in Anna Netrebko singing Puccini's aria; truth inheres is the observation that the world is not flat; justice inheres in the 13th amendment of the American constitution's abolishment of slavery.

Certainly, there are more goodness-evoking transcendentals such as equality, courage, and honesty that are not included. For this synopsis, justice, happiness, beauty, truth, and justice suffice to argue that goodness is increasing. More justice, more happiness, more beauty, more truth, means goodness is increasing. Of the intrinsic transcendental, happiness can be tied to technology, but it is far more difficult to deal with virtue; its treatment must wait until later.

1. Truth

Although it may not be the absolute truth, truth is good in the sense that it diminishes falsehood, myth, error, misdirection. There are several views as to what constitutes truth, but the only one that makes sense to me is a variation of the "correspondence theory" that holds that *beliefs and statements are true only when they accurately describe reality*. This view establishes truth objectively; that it exists absolutely. But I think we can do without the word "beliefs" because beliefs, unless they are expressed as statements, are subjective and can never be known to anyone other than the person holding them. I might believe the truth of a statement but for it to be absolutely true it must be converted to a statement then proven true apart from my mind. The truth alluded to in the first sentence above is objective reality.

Objective truth is that which is permanently true by definition like the theorems of mathematics such as $(3 \times 4) = 12$, and practically true like the laws of physics such as Newton's law of gravity that is true for most observations but not all. I still hold science to be objectively true in most cases because it fits the correspondence theory of the truth in that it *describes* reality.

Truth is mostly good because knowing or merely relying on it makes life much simpler. Life would be filled with paranoia if we didn't know or couldn't rely on my beliefs to be true. So, the more we know and the more we can rely on knowing is in a way contributing to increasing goodness. However, is it true that all truth is good? Perhaps not! Since the purpose of this synopsis is to demonstrate increasing goodness, I will choose certain forms of truth and assume that those that I neglect have no effect.

Much objective truth can also be based on observation, either directly or as reported by an honest reporter. For example, "James just left the room" or "your blood pressure is normal." There is also much *contingent truth*, such as that written in the history books like the statement "George Washington was the first president of the United States." Much *scientific truth* such as "the earth revolves around the sun" is true and will be true for long periods of time. We also must deal with *probabilistic truth* such as weather forecasts, political polls, the chance that the car coming in the opposite direction on a two-lane highway will not veer into your lane. These forms of truth don't enter into my argument that goodness increases because they are unique, or they don't change. Only the scientific truth is the kind that changes and is useful in my argument that goodness increases.

Mathematical learning is true by definition; science generates learning that is associated directly with what we experience, and science observes, hence it is true by the definition we are working with. Scientific knowledge is practical knowledge that has made the human race better off mainly because of sciences impact on technology. As regards absolute knowledge, science is not willing to take the next step to explore the ground of reality. Instead, science becomes a way for materialists to evoke scientism, the belief that knowledge can only be found through the application of sciences.

Individual minds, using the wonder function part of radial insight, realize noumena. In this way radial insight discovers the possible and in doing so instantiates it. Learning is realized into the world through individual minds and instantiated into the pneuma of the realm of potentiality. Tangential insight makes it possible to spread goodness throughout humanity. When a possibility (both and evil) residing in transcendent reality is actualized, the resulting concept is imprinted on the *pneuma* and becomes part of the noosphere.

Much objective truth is based on observation, either directly or as reported by an honest reporter. For example, "*James just left the room*" or "*your blood pressure is normal.*" There is also much *contingent truth*, such as if what is written in the history books is true, then the statement "George Washington was the first president of the United States" is true. Much *scientific truth* such as "the earth revolves around the sun" is true and will be true for long periods of time and the change is imperceptible. We also must deal with *probabilistic truth* such as weather forecasts, political polls, the chance that the car coming in the opposite direction on a two-lane highway will not veer into your lane. These forms of truth do not enter into my argument that goodness increases because they are one-of-a-kind events, or they do not change.

2. Beauty

In 1757 the philosopher David Hume wrote: "*Beauty is no quality in things themselves: It exists in the mind which contemplates them; and each mind perceives a different beauty.*" (Wikipedia ref. Essay XXXII, Of the Standard of Taste). If beauty exists in the mind, the question is how? A scientist might say that beauty is an emergent property of the brain, but that only kicks the ball down the road to the next question, what is an emergent property? Well, we know what an emergent property is. An emergent property is a description of a common mental experience, and what we know is that, in this case, the common experience is a thing we call beauty. Yes, we know beauty when we experience it. But that's not an explanation. I agree with the part of Hume's assertion that beauty doesn't exist in those things that we judge to be beautiful, but I don't agree that beauty exists in the mind. Where in the mind could beauty possibly exist? Certainly not in the neurons, they belong to the material brain, and what else does science believe exists in the mind? The scientific approach to this dilemma is the contention that all mental experiences are emergent properties of the brain. However, beauty is not a common experience like pain or anxiety, mental states that are stored in the immaterial memory. Beauty is transcendent in that it exists outside the mind as was discussed earlier in <ch.2. D>.

Because the experience of beauty depends on the individual, beauty does not inhere in the thing that we call beautiful. Beauty does not inhere in the photons or sounds that engage our senses. It is the reaction to the beautiful, not beauty, which is stored in memory. Beauty must exist in an individual's mind as an emotion. Hence, I define beauty as follows: *beauty is the human reaction to things that are beautiful.* I know that reads like a circular argument except for that phrase "human reaction" that separates beauty from beautiful. We say that 'beautiful' describes things that exhibit 'beauty' and that is the complete definition of the word beautiful.

We all have had the common experience of feeling pain that describes what I mean by *human reaction*. Pain is like beauty in that it is a common experience, but pain is subjective, and beauty is not. Science contends that both pain and beauty are emergent properties of the brain and appear to be stored somewhere in a mind. That is a true descriptive observation, but it explains nothing.

Science knows not where beauty is stored because science recognizes only the material part (the brain) of the mind. My view, which begins with the premise that God exists, allows me to describe an immaterial memory in which subjective experiences such as sensations and feelings are stored. However, pain is stored but beauty is not. Pain is connected directly to the sense receptors; beauty is not, it must be experienced as an emotion that resonates with certain noumena that exist in the pneuma of the realm of potentiality. We experience beauty indirectly through our senses because when beauty is sensed, it is not stored in the brain or in the immaterial memory the way qualia are; qualia such as redness are experienced then realized (brought to mind) as a specific subjective experience as part of a percept that is stored in the immaterial memory.

Natural beauty comes directly from God in sunsets, flowers, the stars, bird calls, waterfalls, and many other naturally occurring sensual experiences. It is what it is, and it occurs when it does. With the exception of those ventures, we take to see a beautiful landmark, most occurrences of natural beauty are spontaneous and can be missed unless we have trained our mind to be aware of such possibilities. I have found that the most effective way to attune my mind to the possibility of beauty is through the faculty of wonder and invariably is accompanied by an emotional feeling of awe. Because natural beauty is a spontaneous happenstance created by God there is no way to link it to a progress principle — God does not progress but is perfectly good eternally — so I have looked elsewhere for support for my increasing goodness argument, to the other modality of called creative beauty.

Created beauty is the beauty created by humans. We humans are made in the image of God, some of whom have been blessed with the capability to create beauty. Although we can never duplicate God's creation of natural beauty, there is much to gain from the beauty of human creation. Creative beauty can be experienced visually, auditorily, and as a mixture of both senses. We can visually experience 2-dimensional paintings, photography, and tapestry; or 3-dimensional architecture, statuary, and pottery. The auditory experience includes music, poetry, and narrations. Both senses engage in the performing arts such as dance, theatre, and motion pictures. And if that isn't enough, we can find beauty many other non-aesthetic endeavors such as clothing, jewelry, and machines. Unlike natural beauty that is sporadic and must be experienced when it is encountered, or it is lost, much of created beauty is semi-permanent and can be apprehended at three levels of interest. Creative beauty can be enjoyed, admired, and appreciated. To *enjoy* the beautiful means to apprehend solely the pleasure that it affords. One apprehends the beautiful with wonder and is rewarded with an enjoyable feeling of awe. To *admire* the beautiful means to discern the art for the quality of the work. One engages the beautiful with discernment and is rewarded with the feeling of amazement. To *appreciate* the beautiful means to scrutinize both the art and the artist for the inspiration it engenders.

Unlike natural beauty that is transitory, creative beauty persists, so that we can apprehend Egyptian Art, The Taj Mahal, Leonardo's Last Supper, Beethoven's piano sonatas, Liszt 2nd, Romanian Rhapsody. We can listen to Marie Callas, with the small handheld cell phone and a touch of the screen at small section named YouTube. With the modern unbelievable store of beauty and information called the internet, the contention that goodness is increasing is one way to demonstrate technology's linkage to beauty.

3. Happiness

Unlike truth, beauty, and justice that are derived from external occurrences such as beautiful objects or human relationships, happiness, like beauty, is a human reaction, but unlike the reaction we call beauty that is induced externally, happiness is induced internally. Whereas the beauty reaction is a transitory emotional feeling, the happiness is a relatively persistent emotional state. Happiness depends on two factors: (i) fulfillment of one's basic needs and (ii) one's personality. For the most part, humans can find happiness only when they have enough to eat, are clothed and sheltered from the elements, have comfort and when all the basic instincts have been met. However, there are those that are born with a resilient personality that appears to be happy in any set of conditions. Such people have a low level of fulfillment of basic needs to have some degree of happiness. Nevertheless, there is some level at which the most resilient person becomes unhappy. The opposite is also true, there are people that are never happy however richly their needs are met. In any case, sustenance is the fundamental need for happiness. From my own direct experience, based on intimate encounters with my eight children and twenty-two grandchildren, each person is born with an innate and individual personality. Some people are born happy while for others nothing seems to make them happy. However, despite the innate component, the personality also has a conditioned component, so that person that were born happy if exposed to negative conditions can become an unhappy person. To argue happiness contributes to increasing goodness, I can only contemplate the level of satisfaction of an average person, the only one I am familiar with; that would be me. To speculate about the satisfaction levels of the basic instincts we need to recognize that each one of them, identity, security, and stimulation are best manifested by a spectrum of satisfying elements,

The basic needs are identity, security, and stimulation. The question that must be answered is at what level of the need for identity, security, and stimulation must be fulfilled for a person to be happy. The simple answer is that it depends on how well one is accepted in the social unit to which a person belongs. The social unit might be a family or a group of friends, co-workers, fellow students, a team or any other such grouping. Happiness is difficult for those that are rejected.

4. Justice

Like beauty, justice is transcendental, meaning it is not stored in the immaterial memory. Unlike beauty that is associated with both objects and actions, justice inheres only in actions among humans. Justice can be experienced individually or as group and we can deal with both individual justice and social justice. Either way, justice is a measure as how people are treated with liberty, equality and fairness either individually or socially. Consequently, justice depends on polity, the way humans organize as social units.

Liberty, the third principle of justice, unlike equality and fairness is derived. Social justice is present when the polity allows, and better yet encourages, the individual to follow the natural law as defined by Judeo Christian principles. There are many social units in which a person can be treated with equality, but without liberty. For example, military organizations might treat each member with equality, but not with liberty. Liberty is the freedom to do as one pleases without impinging on another's freedom. Justice in organizations that depend on hierarchal responsibility exists if the rules for equal treatment are applied equally to all members at each hierarchal level. In other words, there should be no "singling out" of individuals to deprive or enhance their liberty. On the other hand, justice can exist without equality when there is a hierarchy of responsibility.

Equality means that each person is entitled to certain natural rights such as those associated with fairness and liberty but also the most basic right to life and a share of the means of sustenance, to be fed, clothed, sheltered at a basic level shared by the other members of one's social unit. Whereas equality deals with the basic instinct of sustenance, fairness and liberty deal with the basic instinct of identity within each social unit. Social units, from the family to the nation, generally exist with a hierarchy of responsibilities, which generate a commensurate system of privileges. Privilege is the antithesis of equality however justice can still exist if the necessary inequality is treated with fairness.

Fairness means that the equal should be treated equally and the unequal (having more responsibility) shall be treated unequally (having more privilege) contingent, of course, on the principle of equality is fulfilled first. *The king's granary shall be emptied before the first peasant dies of hunger.*

Chapter 26—Charity

*"And now there remain faith, hope, and charity, these three: but the greatest of these is Charity." —*1 Corinthians 13:13 NIV

Charity, in Christian thought, the highest form of love, signifying the reciprocal love between God and man, is made manifest in unselfish love of one's fellow men. The essence of Christianity is love, the essence of love is sacrifice, and all sacrifice is an act of charity. This joins love and charity, two words often interchanged as a theological virtue. Where charity is pure sacrifice, love is more; love compels us not only to sacrifice, but to do good simply for the joy of giving. In a previous chapter, I proposed a purpose for creating humans, namely, to eliminate evil, but that was only half our responsibility. The other half was to choose between doing good or doing nothing.

To understand the incentive for personal good, we need to understand our own human nature. Earlier in discussing rational behavior, I pointed out that the decision to do good or do nothing depended mostly on our internal system of values. The value system is the inner faculty that, when properly acquired, offers options on how to decide when confronted with a choice between positive and negative action. Values can induce not only positive behavior, but through absence can be responsible for also negative behavior, consequently a value system is the most important element in determining positive human behavior because its main purpose is to moderate the basic instincts.

As pointed out earlier, we acquire the value system in three ways, by: inheritance, origin, and education. The first two ways—inheritance and origin—are innate, the last way—through education—we acquire a value system internally through self-development and externally through parenting. Self-development of positive values depends mostly on spiritual behavior that derives from God's grace. However, the strongest development of one's value system is through parenting; and religion is a parent's greatest help for the job.

Chapter 27—Spiritual Reality

Like the other five stages, Noosgenesis consists of a seminal event, the creation era, and the objective. The other creation events were followed by an extended stasis period of slow growth. For example, the big bang was a seminal event that lasted about 380-thousand years before stable hydrogen and helium formed and began the work of creating stars and galaxies. It took 100-million years before the first star began to shine. It was at that point that we can say that the universe was created. Hence, the seminal event lasted 380-thousand years, and the creation era lasted 100-million years at which time the objective (the universe) was reached, and the cosmological stasis period began. The spiritual stage differs from the first five stages in that it is not yet complete: we are in the midst of the era of creation for which the objective has not been reached or even known. I merely assume that the objective is the Omega point, the creation of world peace, and the end of ignorance and oppression.

This creation era, in which we find ourselves, is characterized by an increasing goodness that implies that reality was created with a purpose, a view that clashes with the positivist philosophy that there is no purpose in life, that life is nothing more than the result of a random interaction of chemicals. The meaning I have found in life, on the other hand, is fraught with hope built on purpose. I argue that the purpose of human life can be ascertained from the increasing goodness in the world, an increasing goodness that is a thinly veiled characteristic of human history.

A. Religion

I use the term religion to mean (i) a belief in God; (ii) a set of formalized beliefs; (iii) shared by congregations of persons; (iv) and gathered to perform prescribed rituals dedicated to God.

I am a Roman Catholic; it is the only religion with which I am familiar. But this is a polemic, an argument against secular humanists.

I was born into a non-practicing Catholic family and did not become a practicing Catholic until I married my wonderful wife. It was through the sacrament of matrimony that I learned the meaning of grace and the value of Catholicism and the beauty of the sacraments. It was through the application of wonder that I learned Christianity. I am a Catholic and try to be a Christian; being Catholic and Christian is not always the same thing, although the purpose of Catholicism is to create Christians. It may also be true that one does not have to be Catholic or any other religion to be justified. Justification is achieved through adherence to a Judeo-Christian ethic.

> *"I believe in God, the Father Almighty, creator of Heaven and Earth: and in Jesus Christ His son, our lord, who was conceived by the Holy Spirit, born of the Virgin Mary, suffered under Pontius Pilate was crucified, died, and was buried. He rose again, and descended into Heaven, from which He shall come to judge the living and the dead. I believe in the Holy Spirit, the Holy Catholic Church, the forgiveness of sins, the resurrection of the body and life ever after lasting, Amen."*

In that short paragraph I proclaim my belief in the religious mysteries: God, Trinity, God-centered creation, Virgin Birth, Resurrection, Catholicism, Penance, the resurrection, and Heaven. God exists, so there is hope, and since there is hope, there is a purpose and finding it has been my goal. God created us and must have had a reason. By shining light of reason on creation, I found a way to explain how God "might" create and sustain, and from that derived the nature of hope and life's purpose in the answer to the question: why are we here?

What I have written describes what religion does for me personally, but there is another reason for bringing religion into one's life and that is that it is an essential aid in raising a family. Practicing religion brings: (i) a sense of togetherness to a family, and (ii) tool for providing a righteous value system in one's children.

B. Ritual

"Ritual allows those who cannot themselves out of the secular to perform the spiritual, as dancing allows the tongue-tied man a ceremony of love." — Andre Dubus: A Father's Story in "Times Are Never So Bad"

As the ritual of dancing allows two to become one; church ritual allows disparate minds to join as a single mind. Ritual is at the core of Catholicism; the binding substance that has allowed it to survive two thousand plus years of human turmoil. Ritual is a method for sharing communal thoughts and being rewarded with a feeling of togetherness such as the feeling of joy one gets when being part of a group of fans when our team wins the game. Ritual is the reward for experiencing identity and stimulation together. And even when the opposite happens, and your team loses the fulfillment of one's social identity mitigates the effect one's need of stimulation becomes solace.

We experience joy when we practice charity by the sacrifice of our needs for that of another. Sacrifice is the psychical state temporarily infused by God to enlighten the mind or strengthen the will to perform actions that increase one's justification; it appears in us as a mental state signified as *love*. Love induces a wide variety of behaviors, the main characteristic of which is sacrifice. True sacrifice rewards us with a feeling of *joy*. *Sufficient Grace* is Actual Grace that we may or may not accept. *Efficient or Efficacious Grace* is Actual Grace accepted and acted upon through sacrifice and rewarded with a feeling of joy. *Sanating Grace* is the Divine grace in its function of healing, in human nature, the ravages of sin, original and personal.

The Catholic Church teaches that *"faith without works is dead and that works perfect faith."* Actual grace provides the joy we experience when we exercise the virtue of charity.

Compassion leads to action that rewards us with a feeling of joy. By action, I mean more than the formal action associated with community and service. Action to me has a broader connotation; action is the dynamic application of what we learn from wonder and piety. It is putting God's plan in action. It includes sacrifice, consideration, humility, charity, and love. Christ preached just such a way of life and that is why I know He is humanity's personal God. Failure to respond to actual grace results in feelings of despair.

When loved one-way action generates a false grace that we call pride. Joy is present in the action of love given to another and enhanced when love is mutual and triply enhanced when love is communal.

Sacrifice is a word we think of to describe acts of bravery: the soldier jumping on a grenade to save the life of others; a firefighter entering a burning building to save a child;

Christ on the cross. But sacrifice does not have to be of the dramatic kind; it can be associated with simple acts of everyday life: giving up one's place in line; visiting the elderly in an old people's home; or expressing any other small gesture for example: (i) to part with time and treasure, (ii) to resign oneself to a sacrificial offer; (iii) to offer your love. Most sacrifices are associated with the result of giving your time or your possessions. *Charity* is a more specific term than love, a more ambiguous term; it includes an act of sacrifice. However, whereas sacrifice connotes a loss of time or money, charity can be that, but it also has a positive side. It can be associated with feelings and positive feelings that can transform sacrifice into love.

We confess together that good works – a Christian life lived in faith, hope and love – follow justification and are its fruits. When the justified live in Christ and act in the grace they receive, they bring forth, in biblical terms, good fruit. Since Christians struggle against sin their entire lives, this consequence of justification is also for them an obligation they must fulfill. Thus, both Jesus and the apostolic Scriptures admonish Christians to bring forth the works of love." (Section 4.7 no.37)

C. Justification

The basic instincts motivate animal behavior; tangential insight motivates rational behavior; and radial insight motivates spiritual behavior. By *spiritual behavior* I mean following Judeo-Christian ethic summarized in a single sentence, *"The essence of Christianity is love, and the essence of love is sacrifice."* In that simple statement, I pack all the deep understanding I have regarding the Judeo-Christian ethic. The precept is a rule to love not a suggestion. I adhere to that precept completely! But I do occasionally find myself in the flow of grace that indicates that I am following Judeo-Christian precepts. How do I know that I am in the flow of grace, that I am engaging the Judeo-Christian ethic? I have learned to recognize the hand of God when experiencing awe, peace, and joy by accepting and acting on sanctifying, sacramental, or actual grace.

The objectives of spiritual behavior are an increase in virtue and a decrease in iniquity. The functional elements of spiritual behavior that act to achieve the objectives are the theological virtues of faith, hope, and charity. The rewards for behaving spiritually are awe, peace, and joy. If language created the mind and allowed humans to transcend animal nature, it is God's grace that stimulates the soul and allows the rational man to transcend his transgressions and rationalizations.

Spiritual behavior is that which increases the integrity of the soul. To increase the integrity of the soul we must find God in our lives. We find God in three ways: (i) by observation and wonder of God's creation; (ii) by introspection and discernment within our own self; and (iii) by exercising concern and charity for our fellow humans. We find God upward, inward, and outward.

In the immaterial memory, the integrity of the soul is the difference between personhood and selfhood. Selfhood is innate and forms a baseline against which we can measure personhood. Then the integrity of the soul grows as we develop our personhood while suppressing our selfhood. Personhood is a result of how we control our three levels of behavior: animal, rational, and spiritual. The levels of behavior are determined by the faculties and functions that motivate each level. Basic instincts motivate animal behavior; tangential insight motivates rational behavior, and radial insight motivates spiritual behavior. God put us on earth to develop our soul through the recognition and acceptance of God's presence in creation, in ourselves, and in others.

Each person travels their own path of actuality as their life evolves from an animal soul to the rational soul, to the spiritual soul. The three modalities of the soul—animal, rational and spiritual—evolve along a path of righteousness that mirrors humanity's path of actualization. God's grace is the incentive that raises rational behavior to spiritual behavior.

Selfhood is a behavior motivated by basic instincts, the animal in each of us, the source of negative behavior. Selfhood is innate and forms a baseline against which we must compare our personhood., so the formation of the soul is the lifetime development of one's personhood while suppressing the selfhood.

Personhood is a derivative of tangential insight that incites decisions that can direct behavior in either a positive or a negative direction. {**I**}, the symbol on the behavioral charts, represents the sapient faculty. The development of personhood is the result of how we apply the three levels of behavior. We have seen that the first level, behavior motivated by basic instincts, can lead to the negative behavior that forms our selfhood. We can apply tangential insight to modulate the basic instincts to suppress or enhance them. Furthermore, our radial insight can modulate tangential insight through the value component to influence our rational behavior in a positive direction. We know this mechanism as our conscience. Radial insight when properly formed induces {I} in the positive direction.

Rational behavior is a two-edged sword; our volition directs our choices toward either good or evil. Humanity's greatest evils were perpetrated not by dullards but by men of high intelligence. My premise is that God exists, therefore God must have a reason for providing humans with minds and the capability to do wrong. Obviously, God wants us to reject evil and do good, so He gave us rational behavior to recognize evil and good; He gave us free will to make a choice; and He gave us the ability to act on our choices. Obviously, in persons in which tangential insight did not develop well enough to effectively modulate our instincts, God provided another motivating factor, *radial insight* to modulate the sapient faculty, {I}, and to motivate spiritual behavior.

Spiritual motivation works through the availability of *grace.* that, when recognized, induces us to use radial insight, the direct connection between God and humans. Grace is the gift that induces those that accept it, with the opportunity to guide our radial insight towards spiritual behavior. By spiritual behavior I mean that which increases the integrity of the soul, or, at the very least, does not diminish the integrity. I contend that the most important motivational element involved in the formation of a spiritual conscience is the value element of radial insight.

Compared to the basic instincts that motivate animal behavior, grace, the radial insight that motivates spiritual behavior, is a subtle faculty that forms the soul; it is a faculty that *sees* the world through emotions and feelings. Experiencing joy through the practice of charity informs the soul. Grace does not act spontaneously; we must pursue it through the mind's rational search.

Body and mind form the structure with which we seek justification, but it is the soul that provides the impetus that navigates our soul towards justification through emotions. In the immaterial memory chart, I link emotional states to selfhood, our internal person the soul. One can judge the state of the soul by the emotional state of our selfhood. Selfhood is internal; personality is what the world sees. Each of us must judge our own soul. The true person, the reasonable adventurer, is one for whom their personality is a true image of their personhood.

Each one of us is a unique, solitary element of humanity, the molecule of the sapient stasis period. Just as physical elements define the cosmological stage; molecules the geological stage; cells the biological stage; and animals the sentient stage, human minds are the elements of a sapient stage. It is through our soul that we communicate with God, through grace.

In the narrow sense of the word creation, we think of artists, scientists, composers, and other human creators. However, I am using the word *creation* with extremely broad meaning based on the idea of the human as an element of the cosmic s-frame. As such, we are all part of the whole and any contribution to the change in the cosmic s-fame is contributing to the evolving formation of total reality. In other words, by doing nothing more mundane than taking a walk, driving a car, cooking a meal, painting a wall, we are contributing to the evolution of the cosmic s-frame.

Our mundane activities, however, unless they are memorable and become famous do not become actualized and stored in the realm of probability. Only famous creations appear in material objects such as books, recordings, musical scores, buildings, paintings, sculptures, and movies to become part of objective reality.

To Catholics, justification is "*a translation, from that state wherein man is born a child of the first Adam, to the state of grace, and of the adoption of the sons of God, through the second Adam, Jesus Christ, our Savior,*" including the transforming of a sinner from the state of unrighteousness to the state of holiness. This transformation is made possible by accessing the merit of Christ, made available in the atonement, through faith and the sacraments. The Catholic Church teaches that "*faith without works is dead*" and that works perfect faith. In the Council of Trent, which Catholics believe to be infallible, the Catholic Church declared in the VII session in canon IV (countering the view that sacraments are superfluous and therefore to be eschewed as unnecessary) that, "*If any one saith, that the sacraments of the New Law are not necessary unto salvation, but superfluous; and that, without them, or without the desire thereof, men obtain of God, through faith alone, the grace of justification;-though all (the sacraments) are not indeed necessary for every individual; let him be anathema (excommunicated).*"

Chapter 28—Final Thoughts

Anglican John Henry Newman's 1838 Lectures on the Doctrine of Justification (reissued after his conversion to Roman Catholicism) sought to align the Protestant and Catholic understanding of Justification, writing in terms of the Catholic tradition of "Et….et…" (i.e. both…and…) that righteousness was both imputed and infused (he suggested the term "adhere".)

In Part IV, I use the results found in the first three parts to find plausible arguments to deal with the questions that materialists, skeptics, and cynics—the secular humanists—confront theists. We theists normally address non-believer's questions by invoking faith. But faith is a weak weapon with which to confront: (i) the materialist that invokes scientism; (ii) the skeptic that invokes the absence of proof; and (iii) the cynic that evokes the existence of evil? To enhance faith and to support the belief of the heart with the belief of the mind. I utilize the other two theological virtues—hope and charity— to support faith.

The intent is not to diminish the role of religion as the main defender in the battle for the human mind but instead to arm theists with weapons to counter not only the materialists but also the skeptics and cynics. The scheme for countering these three views of secularism is based on the theological virtues of faith, hope, and charity. It is: (i) with faith that I deal with materialism and skepticism; (ii) with hope that I deal with cynicism; and (iii) with charity to describe the support that faith and hope.

From the first gaze into the starry sky of my youth to the final reflections in these pages, I have journeyed through many questions some of which are scientific, or philosophical; however, all are spiritual. What began with wonder matured into a quest, that led me through the terrain of objective reality, subjective awareness, rational reflection, and spiritual fulfillment. I have not offered a definitive proof, but a plausible alternative; a model of dual reality that integrates science and faith, reason and revelation.

If we accept that our existence is not an accident of chemistry, but the outgrowth of divine purpose, then our lives are imbued with eternal significance. The model I propose does not conflict with the triumphs of science; rather, it seeks to explain the fullness of reality, including those elements science cannot quantify: consciousness, love, beauty, and meaning.

To live is to confront mystery. To think is to seek understanding. But to hope, to have faith, and to practice charity are the hallmarks of a life lived not merely in the physical world, but within the mind of God. It is here that we transcend the animal, elevate the rational, and embrace the spiritual.

We are called not only to understand, but to become fully human, bearing both the image and purpose of our Creator. We are not passive observers in a random universe. We are participants in the Divine, co-authors of our soul's specific story, and recipients of a grace that invites us into eternal communion.

The meaning of life is not a riddle to be solved, but a relationship to be lived. To know God, to love others, to serve humbly, and to hope boldly is our purpose. In pursuing that purpose with eyes wide open and hearts full of wonder, we begin to live not just momentarily in time, but eternally in grace.

So, as the stars continue to shine silently in the sky, I offer this final thought: the universe may be beautiful and vast, but the love that called it into being is greater still. And in that love, we find our origin, our meaning, and our destiny.

—*Frederick Joseph Koons, 2025*

CITATIONS

<1> Explicative level

1. Barr, Stephan M.

"It sometimes makes sense to ask "how" something works, in the sense of seeking a 'mechanism' by which it happens, but sometimes it does not. For example, one can ask the question of how a phonograph produces sound, or how a television produces a picture. But it is not clear that it makes sense to ask "how" a mass produces a gravitational field, say. In Newton theory "mass" and "gravitational field" are fundamental concepts. Newton's law of gravitation posits the existence of a relationship among them and gives a quantitative account of that relationship. But it does not explain "how" the mass produces the field. As Newton himself said in the concluding words of his Principia: I have not been able to discover the cause of those properties of gravity from phenomena, and I frame no hypotheses...[It] is enough that gravity does really exist and act according to the laws that have explained and abundantly serves to account for all the motions of the celestial bodies, and our sea. Similarly, Einstein's theory, while it gives deeper understanding of what a gravitational field is, namely the curving of space-time, does not explain 'how' a massive body causes that curvature, in the sense of a mechanism. **Modern Physics and Ancient Faith**, University of Notre Dame Press, 2013 page 206

2. Peat, David F.

"The world of explicate structures and sequential processes in time, which has been studied by science over the last centuries, now turns out to be a manifestation of a deeper, enfolded order that constantly sustains them." —***Synchronicity***, page 185

<2> Infinite Nothingness

3. Greek Philosophers

It is not known when the first humans wondered about the nature and the beginning of reality, but we know that the presocratic Greek philosophers certainly did. The Greek philosophers believed the beginning or ultimate reality (arche) is eternal and infinite, or boundless (apeiron). The Jewish Kabbalah called it *Einsof* meaning *unending* and meaning *no end* or *infinite*. According to kabbalistic teachings, before the universe was created there was only Ayin. Chinese philosophers called it *Wuji*, a cosmological term for the *Supreme Ultimate* a state of undifferentiated, absolute, and infinite.

4. Rees, Martin

"Cosmologists sometimes claim that the universe can arise 'from nothing.' But they should watch their language, especially when addressing philosophers. We have realized ever since Einstein that empty space can have a structure such that it can be warped and distorted. Even if shrunk down to a point, it is latent with particles and forces—still a far richer construct than the philosopher's nothing. Theorists may, someday, be able to write down fundamental equations governing physical reality. But physics can never explain what breathes fire into the equations and actualized them into a real cosmos. The fundamental question of 'why is there something rather than nothing?' remains the province of philosophers." —***Just Six Numbers***, Page 145

<3> Space

5. Wolfram, Stephen

"*But if the ultimate model for the universe is to be as simple as possible, then it seems much more plausible that both space and its contents should somehow be made of the same stuff—so that in a sense space becomes the only thing in the universe.*" — *A New Kind of Science*, page 474

6. Hoffman, Ernst aka Lama Anagarika Govinda

"*The fundamental element of the cosmos is Space. Space is the all-embracing principle of higher unity. Nothing can exist without Space. ... According to ancient Indian tradition the Universe reveals itself in two fundamental properties: as Motion and as that in which motion takes place, namely Space. This Space is called Akasha derived from the root kas, 'to radiate, to shine', and has therefore the meaning of ether which is conceived as the medium of movement. The principle of movement, however, is Prana, the breath of life, the all-powerful, all-pervading rhythm of the universe.*" —1969

<4> Discrete Space

7. Chaitin, Gregory

"*Nevertheless, there are some intriguing hints that this universe may in fact be a discrete digital universe, not a continuous analog universe the way most people would expect. In fact, these ideas go back to Democritus, who argues that matter must be discrete, and to Zeno, who even had the audacity to suggest that continuous space and time were self-contradictory impossibilities. Through the years I have noticed many times, as an armchair physicist, places where physical calculations diverge to infinity at extremely small distances. Physicists are adept at not asking the wrong question, one that gives an infinite answer. But I am a mathematician, and each time I wonder if Nature was not really trying to tell us something, that the real numbers and continuity are a sham, and that infinitesimal small distances do not exist!*" —*Meta Math, The Quest for Omega*, pages 91, 92

8. Dedekind, Richard

"*If physical space has at all a real existence, it is not necessary for it to be continuous; many of its properties would remain the same even if it were discontinuous. And if we knew for certain that physical space was discontinuous there would be nothing to prevent us, in case we were so desired, from filling up its gaps, in thought, and thus making it continuous; this filling up would consist of the creation of new point-individuals and would have to be affected in accordance with the above principle.*" —*World of Mathematics*, page 530

9. Davies, Paul

"*If the universe is a giant computer as Frank Tipler, the author of The Physics of Immortality, contends then space is surely discrete. A computer operates on bits of information, i.e., it is digitized. For the universe to be computerized its elements— matter, energy, space, and time—must be digitized to allow isomorphic mapping. One objection to Tipler's theory is that the theories of physics are different from computer programs because computers model digitally and physics models with continuity.*" — *Mind of God*, Touchstone (1992)

10. Aristotle

"the void exists...It is the void that keeps things distinct, being a kind of separation and division of things. This is true primarily of numbers, for the void keeps them distinct." —quoted in J. Robinson, **An Introduction in Greek Philosophy** (1968) Boston, page 75; quoted in John D. Barrow, **The Book of Nothing** (2000), page 63

"Aristotle's demonstration of what has come to be known as the isomorphism thesis, which asserts that either magnitude, time and motion are all continuous, or they are all discrete." —**Stanford Philo Encyclopedia**, insert in Zeno's section.

11. Einstein, Albert

"All these fifty years of conscious brooding have brought me no nearer to the answer to the question, 'What are light quanta?' Nowadays every Tom, Dick and Harry thinks he knows it, but he is mistaken. I consider it quite possible that physics cannot be based on the field concept, i.e., on continuous structures. In that case." —Albert Einstein, (1954)

<5> Hylomorphism (Spatial Dualism)

12. Teilhard De Chardin, Pierre

"It is impossible to deny that deep within us an 'interior' appears at the heart of beings, as it were seen through a rent. This is enough to ensure that, to one degree or another, this 'interior' should obtrude itself as existing everywhere in nature from all time. Since the stuff of the universe has an inner aspect at one point of itself, there is necessarily a double aspect to its structure, that is to say, in every region of space and time—in the same way, for instance, as it is granular: "co-extensive with their Without, there is a Within to things." —**The Phenomenon of Man**, page 56

13. My Thought

The sandcastles that we make at the beach provide an analogy for this substance dualism. We can only make sandcastles with wet sand. Water, an analogy for pneuma, is amorphous (not formative), and the sand, the analog for matter, is discrete (is formative). Water forms the sand, and sand forms the water just as the soul informs the body and the body informs the soul. Although the sandcastles are not a perfect analogy, they serve to convey the concept of the substance dualism form of hylomorphism.

<6> Pneuma

14. Heisenberg, Werner

"In the experiments about atomic events we have to deal with things and facts, with phenomena that are just as real as any phenomena in daily life. But atoms and the elementary particles themselves are not as real; they form a world of potentialities or possibilities rather than one of things or facts. The probability wave means tendency for something. It is a quantitative version of the old concept of potentia from Aristotle's philosophy. It introduces something standing in the middle between the idea of an event and the actual event, a strange kind of physical reality just in the middle between possibility and reality." —**Physics and Philosophy,** Werner Heisenberg (1959), page160

15. Barrow, John D.

"In complete contrast with the Atomists' dogma the Stoics believed all things were a continuum bound together by a spirit—an elastic mixture of fire and air—or pneuma, that permeated everything. No empty space could exist within or between the component pieces

of the world, but this did not mean that there could not exist any empty space at all. Quite the contrary, the Stoics Universe was a finite continuous island of material diffused by pneuma but sitting in an infinite empty space. The void was the great beyond and the pneuma bound the constituents of the world together to prevent them diffusing out into the formless void." —*The Book of Nothing,* John Barrow, Vintage Books, New York page 63, quoted from *Physics of the Stoics,* S. Simbursky, London, 1987

16. Einstein, Albert
"According to the general theory of relativity space without **ether** is unthinkable; for in such space there not only would be no propagation of light, but also no possibility of existence for standards of space and time. But this **ether** may not be thought of as endowed with the quality characteristic of matter, as consisting of parts (particles) which may be pathed through time." *Leiden Lecture,* 1928

17. Greek Philosophers
In opposition to the atomists, the Stoic philosophers Zeno of Cition (fl. 250 BCE) and Chrysippus (280–206 BCE) upheld the Aristotelian position that space, time, matter, and motion are all continuous (Sambursky 1954 [1956], 1959; White 1992). And, like Aristotle, they explicitly rejected any existence of void within the cosmos. A continuous invisible substance which they called *pneuma* (Greek: *breath*) filled any void. This pneuma— regarded as a kind of synthesis of air and fire, two of the four basic elements, the others being earth and water—was conceived as being an elastic medium through which impulses are transmitted by wave motion. All physical occurrences were viewed as being linked through tensile forces in the pneuma and matter itself was held to derive its qualities from the *binding* properties of the pneuma it contains. —*Stanford Encyclopedia of Philosophy*

<7> Holism (Cosmic S-Frame)

18. Bohm, David
"...man's general way of thinking of the totality, i.e., his general world view, is crucial for overall order of the human mind itself. If he thinks of the totality as constituted as independent fragments, then that is how his mind will tend to operate, but if he can include everything coherently and harmoniously in an overall whole that is undivided, unbroken and without border (for every border is a division or break) then his mind will tend to move in a similar way, and from this will flow an orderly action within the whole." —*Wholeness and the Implicate Order,* 1980

19. Einstein, Albert
"When forced to summarize the general theory of relativity in one sentence: Time and space and gravitation have no separate existence from matter." —*Ideas and Opinions,* 1954

20. Leibniz, Gottfried
"Reality cannot be found except in One single source, because of the interconnection of all things with one another." —*Discourse on Metaphysics,* (1686)

<8> Incrementation

21. Davies, Paul

"On the other hand, the continuity of space and time are only assumptions about the world. They cannot be proved, because we can never be sure that at some small scale of size, well below what can be observed, space and time might not be discrete. What would this mean? For one thing it would mean that time advanced in little hops, as in a cellular automaton, rather than smoothly. The situation would resemble a movie film which advances one frame at a time. The film appears to us to be continuous because we cannot resolve the short time intervals between frames. Similarly, in physics, our current experiments can measure intervals of time as short as 10^{-26} seconds; there are no signs of any jumps at that level. But however fine our resolution becomes, there is still the possibility that the little hops are smaller. Similar remarks apply to the assumed continuity of space." —**Mind Of God,** page124

22. Spinoza

"We are a part of Nature as a whole whose order we follow." —**Ethics,** 1673

<9> Divine Impetus

23. Aristotle

"Aristotle's term...this God is the Unmoved Mover or Prime Mover, since He presides over everything that chances or moves in the universe, without changing or moving Himself. He alone has achieved His actuality, or energeia, simply by being. He thinks, and everything moves."

<10> Spatial Matter

24. Einstein, Albert

"Physical objects are not in space, but these objects are spatially extended (as fields). In this way the concept of 'empty space' loses its meaning. The field thus becomes an irreducible element of physical description, irreducible in the same sense as the concept of matter (particles) in the theory of Newton. The physical reality of space is represented by a field whose components are continuous functions of four independent variables - the co-ordinates of space and time. Since the theory of General Implies the representation of physical reality by a continuous field, the concept of particles or material points cannot play a fundamental part, nor can the concept of motion. The particle can only appear as a limited region in space in which the field strength or the energy density are particularly high." —**Metaphysics of Relativity,** 1950

25. Einstein, Albert

He [Einstein] argued that if one believed wholeheartedly in the basic idea of field theory, matter should enter not as an interloper but of an honest part of the field itself," recalled one of his Princeton collaborators, Banesh Hoffman. "Indeed, one might say he wanted to build matter out of nothing but convolutions of space-time." —**Einstein,** Walter Isaacson, page 512

26. Einstein, Albert
"When forced to summarize the general theory of relativity in one sentence: Time and space and gravitation have no separate existence from matter." —**Ideas and Opinions,** 1954

27. Schrödinger, Erwin
"What we observe as material bodies and forces are nothing but shapes and variations in the structure of space. Particles are just schaumkommen (appearances)." —**Life and Thought,** 1989

<11> Incremental Time

28. Gödel, Kurt
Commenting on the theory of relativity: *"The existence of an objective lapse of time", he wrote "means that reality consists of an infinity of layers of 'now' which come into existence successively. But if simultaneity is something relative, each observer has his own set of 'nows,' and none of these various layers can claim the prerogative of representing the objective lapse of time."* —Einstein, Walter Isaacson, pages 510–11)

<12> Qualia

28. Edwin Schrodinger
"The sensation of color cannot be accounted for by the physicist's objective picture of light-waves. Could the physiologist account for it if he had fuller knowledge than he has of the processes in the retina and the nervous processes set up by them in the optical nerve bundles and in the brain? I do not think so." —**What is Life? The physical aspects of the living cell,** Erwin Schrödinger, 1958, Cambridge, UK: Cambridge University Press.

<13> Basic Instincts

29. Ardrey, Robert
"Identity, stimulation, security: again, if you will think of them in terms of their opposites their images will be sharpened. Identity is the opposite of anonymity. Stimulation is the opposite of boredom. Security is the opposite of anxiety. We shun anonymity, dread boredom, and seek to dispel anxiety. We grasp identification, yearn for stimulation, conserve, and gain security. And brood as I may over Janus' three faces I have yet to discover a fourth." The Territorial Imperative, page 334

<14> Tangential Insight

30. von Helmholtz, Hermann
The great German physicist, speaking in 1891 at a banquet on his seventieth birthday, describes the way in which his most important new thoughts had come to him. He said that after previous investigation of the problem: *"in all directions... happy ideas come unexpectedly without effort, like an inspiration. They have never come to me when my mind was fatigued, or when I was at my working table... they came particularly readily during the slow ascent of wooded hills on a sunny day."* —From an essay **The Four Stages of Thought** by Graham Wallace in **Toward Liberal Education**

31. Penrose, Roger

"I imagine whenever the mind perceives a mathematical idea it contacts Plato's world of mathematical concepts... When one "sees" a mathematical truth, one's consciousness breaks through into this world of idea, and makes direct contact with it... When mathematicians communicate, this is made possible by each one having a direct route to truth, the consciousness of each being in position to perceive a mathematical truth directly, through this process of 'seeing.' Since each can contact Plato's world directly, they can more readily communicate with each other than one might have expected. The mental images that each one has, when making this Platonic contact, might be different in each case, but communication is possible because each is directly in contact with the same eternally existing Platonic world!" —**Mind of God**, Davies, page 144

<15> Ritual

32. Dubuis, Andre

"Ritual allows those who cannot themselves out of the secular to perform the psychical, as dancing allows the tongue-tied man a ceremony of love." —**A Father's Story in 'Times Are Never So Bad,** loc 2899

33. Mumford, Lewis

"Pre-man must have had dreams, both creative and destructive, a source of disorder and fear. To counteract this, pre-language man communicated through body gestures. Because of its own organic reward, it became rhythmic and communal, constantly repeated until its form with its meaning became sacred and became ritual. Ritual provides order to an otherwise disordered life and formed the bridge between man and animal."

GLOSSARY

abstraction: a term that represents a statement that describes an observation, phenomenon, hypothesis, theory or a deeper reality.

actualization: the creation of reality: objective reality from God through

affection: a positive emotion of feeling such as tenderness

afferent: the direction of information flows in nerves toward the brain

afferent event: the interface of body and mind during which afferent neurons activate cortical neurons to induce subjective experiences such as qualia, feelings, and emotions in the immaterial memory.

afferent paths: the part of the nervous system that carries sense information to the central nervous system, where the information is processed.

Akashic-field: "*Akasha*" an ancient Sanskrit word for space, is used by Lazlo to describe the fundamental energy and information carrying field that forms all universes past and present, it is not exact but like my use of the word "infinite nothingness."

alpha particles: consist of 2-protons and 2-nuclei, like the helium nucleus.

amorphous: without form

analog: an adjective that describes entities that have a continuous nature; it distinguishes *digital* entities that have a discrete nature.

ancillary: subordinate, or depended upon

before/beyond: that which existed before the big and bang and exists beyond the universe. It is synonymous with realm of possibility, pre-universe

binary fission: the process by which single cells divide to produce two daughter cells.

binding problem: the mystery of how the brain combines mental elements such as shape with qualia to form a mental image of, for example, a red apple, when the redness is formed in one part of the brain and the shape in another part.

bios: the psychical substance that is contained in and animates cells.

black-body radiation: a radiating entity in which its surface is in thermodynamic equilibrium; the surface has perfect absorptivity at all wavelengths. It is also a perfect emitter.

body-mind problem: the body-mind problem is science's inability to explain how the neurons in the brain create subjective experiences like qualia, feelings, and emotions other than to describe them as emergent properties of the brain. In this synopsis the body-mind problem is explained as the afferent event in which material neurons excite subjective sensation stored in the immaterial memory.

boson: the class of subatomic particles with an integral value (0, 1, 2, ...) spin quantum number.

causal interaction: this is an abstraction for the way that material interacts with the immaterial

centrality: the concept that because every person is infinitely equidistant from the edge of the infinite and since the definition of the center is that it is a point where all radii are equal, and since every person is equidistant from the infinite

cephalization: the concentration of the nervous system, especially the brain, in the cranium.

cephalization scale: the measure of a species' level of intelligence, the level of learning both conceptual and perceptual.

cognition: mental processes that consist of intellect, volition, and will.

cosmological center: is the point determined by the singularity from which all s-points expand.

common view: I use this phrase to combine all areas of scientific inquiry that address the mind-body problem. It includes philosophy, science, and any other that has opinions.

concept: a mental representation consisting of: (i) a signifier—a mental image of, number, musical note or abstraction associated with a neural map stored in the material memory, and (ii) a referent—all the information associated with the signifier stored in the immaterial memory. For example, the word "*fruit*" is a generalization, and "*dark energy*" is an abstraction; generalizations are palpable; abstractions are not.

configured light: the arrangement of photons in light entering the eye

consubstantial inheritance: the combination of the parents' bios during fertilization to form the incipient soul of the child resulting in the inherited blend of the parent's personhood.

corpus: I use this term instead of body to imply it is more than a physical body, instead, it is a body that is corpuspirated, infused with nous.

corpuspiration: I created this word to mean the infusion of the psychical substance into a cellular or a multicellular body.

cosmological center: The point determined by the singularity from which all s-points expand.

coterminous or conterminous: having the same boundary, I use it regarding the amalgam of material and immaterial existing together to form a hylomorphic structure of objective reality

covariant: a situation in which one magnitude varies mathematically with another. For example, the area of a circle is covariant with its radius.

cpg: **c**entral **p**attern **g**enerator is a neural map that produce rhythmic outputs—sans rhythmic inputs—to control activities such as walking, breathing, swimming, chewing.

creationism: a theory that God alone created humans, not Darwinism.

cytonomic mechanism: the incremental reconfiguration of the s-points that form the cellular s-frame controlled by the bios from within.

descriptive level: the level at which the common view is constructed implying it merely describes rather than explains that God creates and sustains at an explicate level.

digital: an adjective associated with entities built from discrete units as opposed to analog that describes things built from continuous elements.

discernment: capability to clearly recognize, analyze, and distinguish ideas and subjective experiences

discrete: to exist in parts separated from its next neighbors; space defined by the rational numbers.

dual model of reality (DMR): the basic model of this thesis, my view of how God might create and sustain reality. It consists of the holonomic mechanism, the hylomorphic structure of reality, separation of levels of operation: (i) the explicate, God's level, and (ii) the descriptive, science's level.

efferent event: the activity that takes place at the synapses when immaterial (nous) enervates matter (neurons) that initiate behavior.

efferent nerves: those that carry impulses from the central nervous system to skeletal muscles.

embryogenesis: the process by which an embryo forms and develops from a fertilized egg (zygote) into a more complex organism with distinct tissues, organs, and body structures.

emergent: a term used to describe phenomena in which an entity formed from lesser entities exhibits properties that the lesser entities don't exhibit, for example, sodium and chorine that are both lethal and combine to form salt that exhibits an emergent property that is not lethal.

emotions: are internal feelings that are short-lived, focused, negative or positive whose source is external or internal and generally have specific causes, for example, the cause of a negative emotion such as anger, jealousy, contempt, and disappointment generally exist external to one's corpus. The cause of negative emotions such as frustration, shame, embarrassment, and regret generally exists in one's mind.

empirical: based solely on experiment and observation.

endemic: prevalent in or restricted to a specific location

endosymbiosis: when a living organism absorbs another living organism.

entanglement: the phenomenon that one particle of paired doublet will affect the other despite the separation. By changing the spin of one instantaneously reverse the twin's spin. It is a consequence of the uncertainty principle.

epiphenomenalism: the view that mental events are caused by psychical events in the brain but have no effects upon any physical events. There are efferent events but no afferent events.

epistemological: that which is known through knowledge in contrast to the ontology that exists objectively

equivalence principle: when mass and/or effects of pairs of observations are equal, for example inertial mass and gravitational mass.

exclusion principle: two or more fermions with same quantum numbers cannot occupy the same space, particle.

explicative: an adjective used in my view to describe events at the ground-level of reality

explicative level: inferring an explanation to contrast with the common view of reality that describes rather than explains at the descriptive level.

faculty: a thing that possessed the means for accomplishing something

field: region of activity discovered by Faraday to allow ease of mathematical description.

formative causation: Phrase used by Rupert Sheldrake to describe the formation of entities that had not previously existed.

forensic: I use it in an argumentative or polemical sense. The ability to present an argument.

gametes: the mature sexual cells, sperm and ovum.

ganglia: are clusters of cell bodies associated with the sensory system.

gene expression: the process by which information from a gene is used to create functional products, usually proteins or RNA molecules

genome: describes the genetic makeup that is unique to each organism. It is why we can identify individuals by their NA in contrast to phenome, the typical genetic makeup of a species.

heliocentric theory: the observation that the planets revolve around the sun.

Higgs fields: a field associated with the creation of mass.

holonomic mechanism: is the mechanism God uses at the explicate level to create and sustain reality; it is a form of cellular automata consisting of input configuration, algorithm, and **impetus**. It is the core of the Dual Model of Reality

holism: the sum is greater than the parts; the cosmic s-frame transcends the s-points from which it is formed.

homo sapien idaltu: an extinct member of the Homo species that was morphologically similar to the Homo sapien. The first man and woman were probably Idaltu.

homo sapien: the human member of the Homo genus.

homeostasis: the self-regulating mechanisms that maintains an optimal internal environment controlled by internal body temperature, tissue fluid, and the sodium, glucose, and water composition of the blood, and other necessary life sustaining factors.

hylomorphic: the integration of material and immaterial substances. Brains are hylomorphic because material neurons are immersed in an immaterial substance called nous.

hypostatic union: Christ is one person with two natures, divine and human in a single body.

Hypostasis: forms of substance underlying different objects: infinite nothingness, pneuma, and nous are three stasis of a psychical substance

{I}: represents the executive faculty in the mind that controls thinking, initiates behavior, and applies free will. I think of it as "Commander I"

infinite nothingness: the substance that preceded our universe; it is absolute being, pure existence infinite in extant, and infinitely divisible and can be modelled as continuous space. When coterminous with our universe it is called pneuma, when corpuspirated in a cell is called bios, and when corpuspirated in a multicellular metazoan is called nous. It can be immaterial, psychical, psychical depending on where it resides. It is the substance of the *Mind of God.*

insight: the capability to apply one's mental functions—intellect, reason, and values to motivate rational decisions.

immaterial memory: memory consisting of percepts and meanings stored in the nous part of the mind.

instantiate: the function by which possibilities in the realm of possibility are transferred to the realm of potentiality to become noumena.

intelligent design: the theory that reality is the work of a higher power; it contrasts the common view that puts forth the statistical impossibility of the common view.

intentional behavior: that which is initiated within the mind by the {I} faculty in contrast to behavior initiated through the sensory system

intentionality: is the mental connection between what is out there and what our mind *intends* it to be, i.e., what we think it to be. What we see as a red apple is in reality a specific arrangement of quarks and electrons, or in my view what is out there is a specific arrangement of s-points.

isomorphism tests: determine whether two mathematical structures (like graphs, groups, or vector spaces) are structurally identical despite appearing different.

Judeo-Christian ethics: set of values and forms of behavior assumed to be shared by both Jews and Christians.

justification: the application of God's grace to transform us from unrighteousness to holiness; to me, it means applying grace and living according to the Judeo-Christion ethic.

language instinct: the hypothesis developed that contends that a language instinct is genetically innate, that it is a sole explanation for the mind and consciousness.

logical positivism: the philosophical principle that defines the role of science and isolates it from metaphysics and religion

materialism: the belief that matter is the only basis for the formation of objective reality

material memory: the interconnected neural map in the metazoan body that stores the information that governs behavior; much is innate, much can be learned,

mathematical formulation: the principle that all subsequent observations must fit a proven equation.

meiosis: cell division in sexually reproducing organisms in which diploid cells are converted to haploid cells

metabolism: the basic process by which cells function and grow; it includes the conversion of food to energy and the use of energy to build materials and perform other biological functions that are important for the growth and replication of the cell.

Metascience: the study and description of the ground of reality manifested at the descriptive level as science.

metazoa: multicellular organisms in which bodies are made of tissues and organs made of differentiated cells. Metazoans are also known as animals.

mind-body problem: finding an explanation of the activation of neurons by psychical cause.

Mind of God: my view of the before/beyond and the infinite nothingness that forms it

mirrored linkage: the connection of a signifier (words, notes) with its signifier image in the nous.

mitosis: cell division in which a parent cell divides into two identical daughter cells.

modality: how a thing is expressed, thus infinite nothingness can be expressed as immaterial, psychical or psychical independently of its modes

monism: the belief that there is only one substance forming reality

morphic resonance: to acquire information directly from the pneuma to create sentience

morphology: the science of the form and the structure of life, both cellular and multicellular.

morphonomic mechanism: the adaptation of the holonomic mechanism to the multicellular organism – also called formative causation

noumena: immaterial entities that exist in the pneuma of the realm of actuality as qualia, percepts, and concepts that are instantiated as rational and subjective reality.

non-locality: a situation in which information is exchanged beyond the distance that light travels.

nothingness: the appearance of a void, the psychical field that is pure being, and absolute existence and active as a psychical field.

nous: a particle of infinite nothingness corpuspirated in the brain of metazoa.

neural maps: the semi-permanent interconnects of neurons devoted to a specific object such as a word or a function such as a muscular movement.

omniscience: having unlimited knowledge

omnipotent: having unlimited power of authority

omnipresent: unlimited presence everywhere

ontology: study of the nature of being or of reality

ontogeny: sequential development of organisms; from an egg to maturity

panpsychism: the idea that the mind (or more recently) forms a background in all of reality. Same as the author's use of the word pneuma

paradigm: is a set of established scientific principles, widely held opinions, and beliefs, which influence how most scientists think and act. A paradigm becomes a collective mindset that presumes to be the arbiter of truth that presents a barrier that any new idea must surmount to becoming part of the scientific worldview.

parallel worlds: the theory derived from the Many Worlds Interpretation of the Schrödinger wave equation.

path-of-actualization (POA): the history of the evolution of material complexity-consciousness.

path of complexity: increase of material complexity

path of incrementation: is the path of actualization in increments the incrementation.

percept: a mental representation stored in the immaterial memory consisting of signifier and a **referent**. The signifier is a symbol, generally a word, stored as a neural map in material memory; the referent is the information and facts associated with a signifier image, stored in immaterial memory

perceptual intellect: the total amount of perceptual knowledge stored in the mind.

perihelion procession: the movement of the point in a planets orbit that is closest to the sun.

personality: the l manifestation of the results of the contest between personhood and selfhood, this is what the world observes and judges about one's behavior and demeanor.

personhood: the results of one's rational motivation, how one uses tangential insight to modulate basic instincts.

phenome: a phenome is a set of observable characteristics and traits that typifies an organism such as skin, hair, and eye color, morphology, and behavior.

physical behavior: the behavior associated with skeletal muscles

plausible: that which may or may not be true; but which seems believable although there is no evidence to prove it true or not true. It is plausibility for which I am aiming. Unlike *faith* for which there is generally no explanation, plausible is always attached to an explanation.

polemic: an extended argument especially involving an opinion or doctrine.

polity: the general expression for a social unit with a system of rules or laws; use it to describe all types of social units organized with specific objectives, rules, and laws written or inwritten

postdictive science: description of phenomena after observation, for example the theory of evolution and much of biological science.

pre-universe: that which existed before the big bang: infinite nothingness, the realm of possibility, and the Mind of God.

progress factors: aspects of reality such as technology that have been observed to have changed positively over time

prokaryotic bacteria: single celled microorganism that has no chlorophyll.

prosody: the patterns of stress, rhythm, intonation, and phrasing in speech or poetry that convey meaning, emotion, or structure beyond the literal words

psychical resonance: a direct connection the mind makes with noumena.

quantum electrodynamics (QED): theory of particle physics that blends quantum mechanics and special relativity to describe how light and matter interact.

quantum jump: sharp upward change in a property such as material complexity.

realm of actuality: objective reality, the material part of the universe

realm of possibility: a way of referring to the Mind of God to emphasize possibility; it is also referred to as infinite nothingness, absolute being, pure existence, the before/beyond, and pre-universe

realize: the process of the human mind connecting with noumena other than t-noumena to create mental representation, connecting with t-noumena is called psychical resonance.

recursive reality: the sequence of parallel universes

referent: consists of the signifier image plus all the pertinent   information related to the signifier. It can be either a concept or a percept.

righteousness: the quality of being morally right, just, or virtuous.

sapience: having a mind, the ability to manipulate symbols, to think; it distinguishes humans from the rest of the animal kingdom.

secular humanism: the grouping of 3-antipsychical beliefs: scientism, skepticism, and cynicism.

selfhood: the innate, subconscious emotional state that arises out of the concept of centrality and is manifested as individuality, self, and ego. It is being the real YOU as you were born. It is the starting point in one's path of righteousness. It diminishes or expands with volition, one one's decisions.

seminal event: the event that introduces a creation event

semantics: the branch of linguistics dealing with meaning; with percepts

sentience: to be alive; the capability of responding to the senses; even when not conscious, even when asleep for example.

s-frame: the 3-dimensional configuration of s-points. It can be the entire cosmos or specific autonomous objects such as cells or multicellular bodies. Thus, we identify a cosmic s-frame, a cellular s-frame and a morphic s-frame.

s-gap: minimum spacing between s-points, which has been increasing since the big bang and is presently equal to the Planck distance.

signal: sensation that has a definite meaning

signifier: the part of a mental representation that is stored in the material memory and is connected to the referent that is stored in the immaterial memory by a signifier image, mostly a word, a phrase, or a specifically identified sensation such as a red light.

signifier image: a mental image of the signifier that is stored in the immaterial memory as part of the percept or concept and communicates with the signifier stored in the material memory in both afferent direction for recognizance efferent events. A word becomes a word image

singularity: the infinitesimal object from which the big bang emerged.

social unit: general expression for how humans are organized for example: family, clan, club, union, society, team, government at all levels, etc.

somatic: a word associated with the nervous system consisting of nerves that connect the brain and spinal cord to voluntary and skeletal muscles.

soul: the internal difference between personhood and selfhood.

space-time continuum: relativity theory combines space and time into a single mathematical object, because the equations of relativity show that both the space and time coordinates of any event must get mixed together by mathematics, in order to accurately describe what we see.

spatial sequence: the path of elements from centrality to selfhood

s-particle: the smallest particle of matter consisting of an accumulation of s-points in the center of an infinitesimal volume devoid of discrete space.

s-point: point of discrete space that can be associated with a rational number from which the pace that defines the universe is composed. Volumes devoid of s-points create materiality.

spontaneous behavior: that which is induced through the sensory system resulting from stimuli in contrast with intentional behavior that is induced by the mind in exercising free will.

stereopsis: depth perception

stochastically: a random process consisting of a sequence of random variables

stromatolites: layered sedimentary deposits formed mainly on limestone by blue-green algae, cyanobacteria and other primitive one-celled organisms.

strong force: the microscopic interaction that binds together particles such as the quarks in a proton and protons in the atomic nucleus.

synopsis: an abbreviation of an extended text. You are reading a synopsis because I minimize citations and extended explanations.

syntax: the arrangement of words in a sentence, grammar.

subjective experiences: those through which each one of us *creates* our world; they include four modes: sensing, feeling, thinking, and behaving. Sensing and feeling are passive modes; thinking and behaving are active modes.

subjective reality: the non-physical world that arises in the human mind.

substance dualism: the view that two distinct kinds of substance produce two kinds of properties.

sustenance: the metazoan faculty responsible for the maintenance of life, its two main functions are homeostasis and metabolism.

symbiotic: the situation of different types living together

telic: having a goal or objective

temporal perceptual intellect (TPI): The ability to perceive and interpret sensory information and apply rational or analytical reasoning to derive meaning, make decisions, or form abstract concepts based on those perceptions.

temporal sequence: the movement of one's body in time, walking or dancing for example

tensor: mathematical object

traducianism: the theory that a person's soul is generated at birth from a contribution from both parents.

transcendental: things beyond objective, subjective, and rational reality. Hence, transcendentals are affections that exist in the realm of potentiality.

transmission problem: explaining how the image of a red apple that is perceived in my mind, appears across the room on a configuration of quarks and electrons.

uncertainty principle: the inability of measuring both a particle's momentum and position with precision; a measure of the particle's momentum would make be uncertain its position by amount called Planck's constant.

verbal image: is the *thought* of a word or phrase that is not spoken or mouthed but rather is *experienced* in the same way as one imagines a face. i.e., as an image.

volition: the power of the executive function to act with insight, motivation, and will.

weak force: is the mechanism that induces the radioactive decay resulting from the interaction between subatomic particles. The weak force participates in nuclear fission and nuclear fusion.

Zeeman effect: when a light source is placed in a magnetic field the interaction of its atoms causes single spectral lines to split into multiple components

zygote: single diploid cell formed when the haploid cells of the sperm and the ovum combine; it consists of a plasma membrane surrounded by a thicker coating called the zone pellucida.

INDEX

225

BIOGRAPHY

I was born on September 3, 1933, in Wilkes-Barre, Pennsylvania, into what I'd call the upper level of the lower class—plenty of capability, not much income. We were poor. I grew up with two sisters and two brothers. In 1943, our family joined the migration from the coal towns to the steel city of Bethlehem, PA. I graduated from Liberty High School in 1951 and worked at several local employers, including Lehigh University's civil engineering lab, Bethlehem Steel, and the Magnetic Windings, testing transformers.

In 1953, I joined the Army and was assigned to the Army Security Agency. I trained in radio repair, then flew to Asmara, Eritrea, with stops in the Azores, Tripoli, Cairo, Dhahran, and Jeddah. I spent two years in Eritrea before returning home and enrolling in the Physics program at Penn State. In June 1958, at the end of my sophomore year, I married a wonderful and beautiful woman. We were both 24. Our first child was born a year later, in June 1959. By taking extra courses, I graduated in just seven semesters and began work in January 1960 as a semiconductor Development Engineer for Western Electric, the manufacturing arm of the Bell System.

Our family grew quickly—eight children in 11 years: a girl in 1959, a boy in 1960, followed by a girl in 1961, a boy in 1962, then girls in 1963, 1965, and 1968, and our youngest, a boy in 1970. The pace slowed a bit after 1962 while I was building our home. We bought a half-acre lot and a pre-cut house. Contractors handled the foundation, framing, roof, brickwork, siding, and cabinets, but I did all the electrical, plumbing, floors, tile, drywall, bathrooms, doors, and trim myself—nights and weekends for 13 months. We moved in, nearly finished, in August 1965 and settled into raising our family.

The '70s and '80s were all about school and college. All eight of our kids made the National Honor Society, and six went on to earn advanced degrees. I stayed active in their lives—serving as a Cub Scoutmaster, Little League coach, CYO basketball coach, and high school science fair judge. In the late '80s, I received a fellowship to earn a Master's in Solid State Physics from Lehigh University.

Our children pursued a wide range of careers. Our youngest daughter, Rose, earned a doctorate in Food Science—yes, that's a real field—and appears on YouTube in the *Epicurious 4 Levels* series, where she explains the science behind cooking techniques. Between 1982 and 2000, seven of our children were married, all in the Roman Catholic Church. Today, we have 22 grandchildren—11 of whom are married—and 23 great-grandchildren (so far). No divorces, and all strong marriages.

I retired in 1995 after 35 years of service. Despite raising a large family, my wife and I made sure our children had the same freedom and opportunities as any American child. We also managed to enjoy a full and active social life—especially in retirement—attending operas and musicals, going on cruises and bus trips, and visiting our far-flung children and extended family across the country. It's been a busy, meaningful, and very rewarding life.